Mining and Metallurgy of the Black Hills of South Dakota

by Black Hills Mining Men's Association

with an introduction by Kerby Jackson

Introduction

It has been years since the Black Hills Mining Men's Assoc. released their important publication "Papers Read Before The Black Hills Mining Men's Association At Their Regularly Monthly Meeting On The Mining and Metallurgy of the Black Hills Ores". First released in 1904, this important volume has now been out of print for this days and has been unavailable to the mining community since those days, with the exception of expensive original collector's copies and poorly produced digital editions.

It has often been said that "*gold is where you find it*", but even beginning prospectors understand that their chances for finding something of value in the earth or in the streams of the Golden West are dramatically increased by going back to those places where gold and other minerals were once mined by our forerunners. Despite this, much of the contemporary information on local mining history that is currently available is mostly a result of mere local folklore and persistent rumors of major strikes, the details and facts of which, have long been distorted. Long gone are the old timers and with them, the days of first hand knowledge of the mines of the area and how they operated. Also long gone are most of their notes, their assay reports, their mine maps and personal scrapbooks, along with most of the surveys and reports that were performed for them by private and government geologists. Even published books such as this one are often retired to the local landfill or backyard burn pile by the descendents of those old timers and disappear at an alarming rate. Despite the fact that we live in the so-called "Information Age" where information is supposedly only the push of a button on a keyboard away, true insight into mining properties remains illusive and hard to come by, even to those of us who seek out this sort of information as if our lives depend upon it. Without this type of information readily available to the average independent miner, there is little hope that our metal mining industry will ever recover.

This important volume and others like it, are being presented in their entirety again, in the hope that the average prospector will no longer stumble through the overgrown hills and the tailing strewn creeks without being well informed enough to have a chance to succeed at his ventures.

Kerby Jackson
Josephine County, Oregon
October 2015

PREFACE.

On September 3rd, 1901, the Black Hills Mining Men's Association was organized, having in view the advancement of the mining industry of the Black Hills of South Dakota, the cultivation of a better feeling among the mining men of the district, and the advertising to the world of the vast mineral resources of what has been described as "the richest 100 miles square in the world."

No definite plan was outlined for this work, but as the association grew in numbers and strength, its field of work developed accordingly. The first thought of its organizers was to get accurate and reliable information about the vast mineral resources of the Hills, the different characters of ore, the methods that were being adopted for the treatment of the same, and to get this information in a concise form for distribution.

About that time the immense bodies of low grade ore found in the Black Hills—and which have caused the Hills to be known as the greatest low grade mining district of the world—began to attract attention. These ores were found to be well adapted to cyanidation and new plants sprung up everywhere. An army of chemists, old and young, became interested at once in this method of treating ores, and set their brains to work to see to what extent the process could be developed in the treatment of the ores. It is conceded that this process is more widely used in the Black Hills than in any other mining district. Several of the papers in this book are upon the cyanide process as it is applied in the different mills of the Black Hills. These, and other papers were read before the association, and created a widespread interest, not only in the process, but also in the meetings of the association. A number of these papers have been published in pamphlet form and distributed all over the world. Calls for them have been received from South Africa, Australia, Central America, and from other mining localities where cyanidation is only in its infancy. It has been thought desirable to publish these papers, together with the others that are contained herein, in one small book, for general distribution, on the ground that it will be more likely to furnish the kind of information desired by mining men. We have added a brief outline of the mining industry of the Black Hills, and also give some statistics showing the output of gold from the Hills and the new reduction plants under headway.

The committee, on behalf of the association, desires to acknowledge its indebtedness to the gentlemen whose papers have been published, for permission to publish them in this form.

MINING AND MILLING IN THE BLACK HILLS.

By Major A. J. and Jesse Simmons.

There are two distinctive features of Black Hills mining that place it apart from any other section on earth. They are the immense bodies of ore and the scientific methods employed in their mining and milling. The extent of these ore bodies can hardly be conjectured. They have been identified at dozens of places, and the limits have many times been established by theorists. But such limits have been as promptly extended by exploration and the opening of new camps. Some writer has said of the Black Hills that "every hill contains a gold mine." The truth of this assertion, while it may sound rather broad, is daily brought home more forcibly upon the report of new discoveries. Had the writer continued his theme one point farther he would have said that "every gulch contained a gold mill."

And it is in these mills that the free-milling and cyaniding processes have been advanced to the highest stage of perfection known to the mining fraternity. The Black Hills challenges the world to compete with its mill practices. It is at the Homestake that ore reduction and amalgamation has been constantly improved for the past quarter of a century until today hydrometallurgists the world over are striving to reach the mark set. In the cyanide mills a revelation has been made to the world in the economical handling of ores. Even South Africa with its Kaffir labor at 25 cents a day cannot treat ores by cyanide as cheaply as the Black Hills, where the standard of wages is $3.00 a day. The Homestake tailings cyanide mills are known and spoken of as remarkable plants wherever mining is conducted. They are remarkable as to size and treatment costs. One has a capacity of 1,800 tons per day and another treats 1,000 tons every twenty-four hours. Costs average 28 cents per ton. The nearest approach to this figure is on the Rand, but there the cost is nearly double, or 53 cents per ton. And in the cyanide mills handling raw ore as it comes from the mines records have been made which have never been approached outside of the Hills. The Deadwood Standard has mined, milled, paid all office, general and incidental expenses at the unprecedented figure of $1.22 per ton. The Wasp No. 2 mill was operated one month on ore assaying $1.60 a ton, and at the end of the month paid a small dividend.

It must not be understood that all Black Hills mines operate as cheaply as the three mentioned, Homestake, Deadwood Standard, and Wasp No. 2. They are peculiarly endowed with physical conditions which tend to greatly lessen operating expenses. In the latter two the ore is virtually quarried, and is of a character allowing coarse crushing and cheap treatment. In the Homestake the enormous size of the ore bodies and the stoping system applied thereto at once make operations cheap, for that peculiar ratio applied to all manufacturing businesses is true of mining, viz: increased output cheapens cost.

Aside from the Homestake, where the recovery averages for years at a time, $3.54 to $5.00 per ton, the average gold values saved per ton is from $5.00 to $8.00. And it is upon the handling of these grades of ore that the Black Hills has built up an industry which places it far in the lead of all mining districts. The enormous bodies of ore, admitting of cheapest mining, and the remarkable adaptation of recovery methods have made the transmutation of wealth from nature's storehouses in this section the marvel of the mining industry.

The following list of Black Hills mills, with their monthly tonnages and treatment methods, gives an idea of the size to which the industry has grown:

STAMP-AMALGAMATION MILLS.

	Monthly Capacity. Tons.
Homestake	120,000
Clover Leaf	7,500
Apex	7,500
Pluma	3,600
Mainstay	3,600
Golden Empire	3,000
Holy Terror	2,400
Montezuma	1,200
St. Elmo	1,200
Clara Belle	1,200
Extreme	1,200
Tykoon	1,200
Sunbeam	1,200
Inca	1,200
Highland	1,200
Golden Slipper	1,200
Cochran Mine	600
Black Eagle	600
Golden West	360

CYANIDATION MILLS.

Horseshoe	15,000
Hidden Fortune	9,000
Spearfish,	7,950
Golden Reward	5,250
Jupiter	4,500
Wasp No. 2	4,450
Deadwood Standard	4,450
Imperial	4,000
Maitland	3,600
Dakota	3,150
Lexington Hill	3,000
Columbus	2,400
Alder Creek	2,400
Lundberg, Dorr & Wilson	2,250
Golden Crest	1,500
Cleopatra	1,500

PYRITIC SMELTERS.	Monthly Capacity. Tons.
✔ Golden Reward	15,000
National	7,500

TIN CONCENTRATORS.

Tinton	3,000
Total	259,860

UNDER CONSTRUCTION.

Branch Mint, cyanide	27,000
Gilt-Edge Maid, cyanide	4,500
Golden West, amalgamation	3,000
Ruberta, amalgamation	1,500
Puritan, concentration	1,500
Grand Total	297,360

These mills are not now all in full commission and probably never were all at any one time—the exigencies of the mining business are the same as that of others. There are occasional stoppages in whole or in part, repairs have to be made, accidents will occur, etc., and probably a fair conservative estimate on the actual annual tonnage of ores treated in the Black Hills mills would approximate about 2,000,000 tons. This is steadily and constantly increasing.

The amalgamation plants are headed by the Homestake, the greatest gold mine in the world, which, with its 1,000 stamps reduces daily 4,000 tons of ore. Following the amalgamation the coarser portions of the tailings are treated by cyanidation.

The Homestake is called the greatest gold mine in the world for several apparent reasons. It has not missed paying a monthly dividend for twenty-five years and seven months, and during that time it and the Caledonia, Deadwood-Terra and DeSmet Companies, all now amalgamated, have paid $20,000,000 in dividends, and produced $90,000,000. It has the greatest annual production of any gold mine, namely, $4,500,000 per annum. It has today in sight in its mines enough ore to supply its present capacity for twenty years, and this at only 1,100 feet depth. Practically no ore has been removed below 800 feet except as taken out in the course of driving development drifts and cross cuts, while there are millions of tons as yet untouched above that level. On the lower levels the ore body is 500 feet wide. From north to south, over two miles, the ore is obtained from numerous open cuts, shafts and tunnels.

All of this has been accomplished upon the handling of ores from which are recovered $3.50 to $5.00 per ton, and the wonder of it is that it has been done. Besides disbursing $20,000,000 to its stockholders, the Homestake Mining Company has earned and expended many other millions of dollars in betterments and expansion of its territory. It has grown from less than ten acres to over 2,600; its milling equipment increased from eighty to

1,000 stamps; its Ellison and other hoists built and equipped at an expense of millions of dollars; its water system from Spearfish River completed at a cost of a round million; all upon the profits of an original investment of $300,000, which paid for the first claim and furnished the first mill and hoist. Thus the Homestake stands alone, a peerless figure in the mining industry, a living monument of solidarity dedicated to the handling of low grade ores on a business basis.

The Homestake Belt of veins, similar to the Mother Lode of California, has been traced through the Hills a distance of forty miles, on a course of south thirty-five degrees east from Maitland, through Lead, and on as far as Keystone, and identified at numerous intervening points. Thus the Homestake has not pre-empted the whole field, but there is abundant room for dozens of like institutions.

The Clover Leaf Mine at Roubaix, on the Homestake Belt, operates a sixty-stamp mill, saving values entirely by amalgamation. Late cleanups show returns of a little better than $6.00 per ton, and at the deepest working, 700 feet, the grade of ore is better than at any point previously opened. Continuing on southward mines are being opened and small mills have been built at frequent points along the Belt, and important developments are today under way at numerous points. At Keystone large low grade veins are identified as Homestake Belt material, while at the Holy Terror Mine an exceptionally rich vein has been milled to a depth of 1,200 feet. The Holy Terror has produced about $2,500,000 and paid $270,000 in dividends. It is equipped with a twenty-stamp mill. Other mills at Keystone are Tykoon and Mainstay.

CONTACT DEPOSITS.

Immediately overlying the Algonkian rocks in which are found the fissures of the Homestake Belt is first a bed of conglomerate, next Cambrian quartzite, then sandstone and shales to a depth of 200 to 300 feet. Intrusions of porphyry, phonolite, trachyte, rhyolite, and other eruptive rocks have twisted and folded the formations to great extent. Mining engineers agree that these eruptions have had great bearing upon the ore deposition, and the miner always looks for porphyry—using the term in the most general sense—in proximity to all ore bodies.

The contact deposits of Bald Mountain, Maitland, Yellow Creek, Two Bit, Blacktail, Galena, etc., have been mined for the past fifteen years, and still new mines and new camps are constantly being opened. When it is considered that about three townships in western Lawrence County are underlaid by this formation, and have scarcely been scratched, the field is open for even greater camps than those at present known. At first the ores were shipped to Omaha, Denver and elsewhere for smelting, the miners utilizing only the high grades. Next matte smelting was introduced to the Hills about 1890, and fifteen to twenty dollar ore profitably handled. Then chlorination made pay ore of ten to fifteen dollar stuff. But it was upon the introduction of the cyanide process that mining of the contact deposits reached the height of its glory. Ores worth from $2.00 to $10.00

per ton are now successfully handled. And it is upon the application of this process to the handling of the siliceous contact ores that the greatest industry of the Hills, aside from the mining of the free-milling belt, is established.

Cyanidation has made available untold millions of dollars contained in these refractory ores. It has undoubtedly done more to place the Black Hills in the front rank of mining communities than any other single factor. And its adaptations as observed in the Hills are studied by mining men the world over. The Black Hills Mining Men's Association is the forum of experimentation and publicity of mill practice. The papers read before it and published in this volume embody data on the practical operation and results of mill superintendents in cyanidation in the district. They may be said to form the classics of our knowledge of cyanidation.

The list of cyanide plants of the Black Hills as previously given includes first, the Horseshoe, the largest complete wet crushing cyanide plant in the world. Its 120 stamps have a capacity of 500 tons of raw ore per day. It is a model in every particular.

Continuing down the list are such mills as Golden Reward, Hidden Fortune, Spearfish, Jupiter, Wasp No. 2, Deadwood Standard, Imperial, Maitland, Dakota, Lexington, Lundberg, Dorr & Wilson, etc., all distinctive in a way, and yet all operating successfully on the great contact deposits of low grade ores.

Spearfish and Deadwood Standard handle ore from the carboniferous lime formation, and during the past two years have paid $123,000 in dividends. Both are plain types of dry, coarse crushing mills. The porous character of the ore admits of coarse crushing and quick extraction is accomplished.

Under the head of tin concentrators the Black Hills offers to the inspection of the world the only producing tin mine on the American continent. The plant has a capacity of 100 tons daily, crushing with rolls and concentrating on Bartlett tables. The cassiterite concentrates, for the time being, are shipped to Swansea for smelting. The vein is in the igneous formation (granite and porphyry) and is disclosed in the workings 100 feet wide, and its out-crop identified for several miles. Adjoining companies are watching operations closely and more plants of a similar character are anticipated, as this mill is opening the road to success.

But after all the Black Hills is a great low grade gold field, where illimitable quantity compensates for high values and resolves mining to a permanent industry—its ore supply will last for ages. The treatment of the immense deposits requires careful management, large milling facilities and economic mining and handling of the ore. The Hills is endowed with physical advantages second to no mining district on earth. Timber, water and railroad communications are par excellence, its climate agreeable, and lastly, it is within easy reach of eastern money centers. One has but to peruse the list of plants under construction—a sample of a year's growth of this wonderful mining district—to form an idea of the future possibilities of the country. Its growth is natural and steady, and one day it will be known as the world's greatest gold producing camp.

SOUTH DAKOTA GOLD PRODUCTION.

According to Hon. Geo. E. Roberts, director of the U. S. Mint, South Dakota stands third on the list of gold producing states.

With the exception of Alaska, an exclusive placer field, South Dakota's annual gold output is only exceeded by the great mining states of Colorado and California.

And when it is considered that the gold producing area of South Dakota is confined to the Black Hills, a section barely forty miles wide by eighty long, the wonderful resources of the country are apparent.

The paraphrase "richest one hundred miles square in the world" fittingly describes this wealthy little mountain range in comparison with the hundreds of thousands of square miles comprising the great mining empires of Montana, Idaho, Oregon or Utah.

The following table gives the total gold output of the Black Hills since the yellow metal was first mined in 1876, up to the year 1903, inclusive:

Year	Total Production
1876	$1,200,000
1877	2,000,000
1878	2,250,000
1879	2,500,000
1880	2,650,000
1881	2,550,000
1882	2,550,000
1883	2,525,000
1884	2,575,000
1885	2,750,000
1886	3,250,000
1887	3,420,000
1888	3,485,000
1889	3,550,000
1890	3,904,160
1891	4,619,270
1892	5,101,630
1893	6,750,000
1894	6,500,000
1895	6,800,000
1896	6,775,000
1897	6,524,760
1898	6,800,000
1899	7,000,000
1900	7,250,000
1901	7,400,000
1902	7,500,000
1903	7,829,000
Total	$128,008,820

SOME FEATURES OF MINING OPERATIONS IN THE HOME-STAKE MINE, LEAD, SOUTH DAKOTA.

By Bruce C. Yates, M. Am. Inst. Min. E.

[Paper read before Black Hills Mining Men's Association, January 19, 1904.]

INTRODUCTION.

The Homestake Mining Co., as it exists today, is a consolidation of the Father De Smet, the Deadwood-Terra, the Caledonia, the Highland and the Homestake Gold Mining Companies. The property comprises some 350 locations with a total area of 2,624 acres.

The ground upon which mining operations are being carried on at present lies between Deadwood Gulch on the north and Whitewood Gulch on the south, covering about 10,000 feet along the strike of the ledge. The principal works and the offices of the company are located almost in the heart of the town of Lead, which has a population of about 8,000. The town is built on either side of Gold Run Gulch, which is a tributary of Whitewood Gulch and which passes through the company's property from west to east a little south of the center.

The strike of the Homestake ledge is approximately north thirty-five degrees west, and south thirty-five degrees east. The dip is very irregular.

The company is now operating 900 stamps and an additional 100 stamps will be installed during the coming year. The 900 stamps are grouped in six mills, the Homestake and Golden Star with 200 stamps each and the Amicus with 140 stamps are located on Gold Run Gulch, the Pocahontas with 160 stamps and the Monroe with 100 stamps are located on Bobtail Gulch which is a tributary of Deadwood Gulch. The Mineral Point Mill with 100 stamps is located on the south side of Deadwood Gulch and near the northerly end of the property. The tailings are treated at two cyanide plants. One with a capacity of 1,450 tons per day is located on Gold Run below the Homestake, Golden Star and Amicus Mills and treats the tailings from these mills. The other, with a daily capacity of 800 tons, is located on Deadwood Gulch at the mouth of Bobtail and receives the tailings from the Pocahontas, Monroe and Mineral Point mills.

The ore is principally free milling, seventy-two per cent. being caught on the plates, the total recovery is approximately eighty-eight per cent. of the assay value. (See paper on Metallurgy of the Homestake Ore by C. W. Merrill, Met. Eng., read before the American Mining Congress at its 1903 meeting in Lead, S. D.)

The main source of the water supply is the head waters of Spearfish Creek. A pumping plant consisting of two Riedler pumps driven by Compound Corliss Engines with a third now being installed is located near the head of the east fork. The water is pumped 400 feet to the top of the divide between Spearfish and Whitewood Creeks from which point it is carried by gravity to the main reservoir at Lead, some ten miles in all.

A very small volume of water is pumped from the mine considering that there are about forty-one miles of track now open. The water was formerly taken care of by a Cornish pump which lifted the water in three stages from the 800 foot level to the surface, until about two years ago when a Riedler Compound Condensing pump, with a capacity of 550 gallons per minute under a head of 1,200 feet, was installed on the 1,100 foot level.

The underground haulage is done by horses and mules. The horses seem to be better adapted to the work than mules and in consequence there are very few mules left in the mine. The cost for horse haulage is three cents per ton per 1,000 feet.

A five ton air motor has been ordered and will be tried on one of the lower levels. If successful, these motors will, no doubt, take the place of horses on all main haulage ways. A fifteen ton motor is now being used on the tramway level to haul ore from the crusher bins to the mills. The air is supplied at a pressure of 950 pounds by a 100 horse power three stage straight line compressor which will serve both motors temporarily.

Following is the tabulated statement of the power required to carry on the different operations:

MACHINE DRILLS.

MANUFACTURER'S NAME	SIZE	NO. IN USE	PURPOSE USED FOR	REMARKS
Ingersoll.	A–32–2¼-in.	48	Block holing.	Six of the Ingersoll are used for Ajax drill sharpening machines and two are used for shop welding in car repair shops.
Ingersoll.	B–32–2½-in.	12	Upper Levels & Surface	
Ingersoll.	D–32–3½-in.	30	General Mining.	
Ingersoll.	D–24–3-in.	148	General Mining.	
Rand.	3½-in.	2	Shaft Sinking.	
Leyner.	3-in.	3	Surface Drifting.	
McKeirman	3-in.	1	Shop Hammer.	
		244		

ELECTRIC MACHINERY.

MNFS. NAME	K. W. OR H. P.	HOW DRIVEN	WHERE USED
Edison	8½ K. W. Gen.	Connected to line shaft	Cyanide No. 2.
Westinghouse	22 K. W. Gen.	Connected to line shaft	Cyanide No. 1.
Edison	6 K. W. Gen.	Driven by 10 H. P. Ideal Engine.	Monroe Mill.
Edison	2 (60) K. W. Gen.	175 H.P. Upright Eng	B. & M. Shaft.
Gen. E. & C. & C.	2 (5) H. P. Motors	Connected to mill sh't	Amicus Mill.
Roth.	2 H. P. Motors	. . . .	Homestake Office.
Edison	2 (20) K. W. Gen.	Line shaft from engine	Golden Prospect Shaft.
Edison	17½ K. W. Gen.	25 H. P. Ideal Engine.	Mineral Pt. Mill.
Brush.	4½ K. W. Gen.	Connected to Mill Eng.	Pocahontas Mill.
Gen. Elec. . . .	4½ K. W. Gen.	10 H. P. Ideal Engine	Pump Station.

HOISTING ENGINES.

MANUFACTURER'S NAME	KIND OF ENGINE	H. P.	WHERE USED
Union Iron Works..	Simple Duplex..........	1200	Ellison Hoist.
Frazer & Chalmers .	Simple Geared Duplex....	150	Golden Star Hoist.
Frazer & Chalmers .	Simple Duplex..........	400	B. & M. Hoist.
Frazer & Chalmers.:	Simple Duplex..........	500	Golden Prospect Hoist.
Frazer & Chalmers..	Simple Geared Duplex....	150	Old Brig.
Frazer & Chalmers..	Simple Geared Duplex....	150	Golden Gate.
		2550	

CRUSHER ENGINES.

MANUFACTURER'S NAME	KIND OF ENGINE	H. P.	WHERE USED
Frazer & Chalmers..	Compound Corliss.......	250	Ellison Hoist.
Frazer & Chalmers..	Simple Corliss..........	250	B. &. M. Hoist.
Frazer & Chalmers..	Simple Engine..........	200	Golden Prospect.
Frazer & Chalmers..	Simple Enigne..........	100	Old Brig.
Frazer & Chalmers..	Simple Engine..........	200	Golden Gate.
		1000	

MILL ENGINES.

MANUFACTURER'S NAME	KIND OF ENGINE	H. P.	WHERE USED
Geo. H. Corliss....	Cross Compound........	500	Homestake Mill.
Geo. H. Corliss....	Tandem Compound......	500	Golden Star Mill.
H. Corliss........	Simple Corliss..........	350	Amicus Mill.
Frazer & Chalmers..	Simple Corliss..........	350	Pocahontas Mill.
Union Iron Works..	2 Simple Engines.......	250	Monroe Mill each 125 H
Union Iron Works..	Simple Engine.........	250	Mineral Point.
		2200	

AIR COMPRESSORS.

MANUFACTURER'S NAME	KIND OF ENGINE	H. P.	WHERE USED
Ingersoll Sargeant..	2 Stage Compound Corliss..	1000	Ellison Hoist Low Pressure.
Ingersoll Sargeant..	3 Stage Straight Line......	100	Ellison Hoist High Pressure.
Ingersoll Sargeant..	Duplex................	500	B. & M. Hoist Low Pressure.
Ingersoll Sargeant..	Duplex................	300	Golden Pros. Hoist L. Pres.
Ingersoll Sargeant..	Single Engine..........	100	Cyanide No. 1 L. Pressure.
Ingersoll Sargeant..	Single Engine........ ...	100	Cyanide No. 2 L. Pressure.
		2100	

PUMPS.

MANUFACTURER'S NAME	KIND OF ENGINE	H. P.	WHERE USED
Riedler............	Compound Corliss........	250	Mine—1100 ft.
Riedler............	2 Compound Corliss......	650	Spearfish Creek each 325 H. P.
1 Holly, 2 Dean....	Compound Corliss........	150	Settling Dam each 50 H. P
Frazer & Chalmers..	Simple Corliss..........	250	B. & M. Shaft Cornish.
3 Prescotts........	Compound...............	120	Cyadine No. 1 each 40 H. P
2 Heisler..........	Triple Expansion........	80	Cyanide No. 2 each 40 H. P.
		1500	

MISCELLANEOUS.

MANUFACTURER'S NAME	KIND OF ENGINE	H. P.	WHERE USED
....	Simple Engine........	25	Ellison Shaft Sinking Eng
....	Simple Engine........	30	Amicus Mill Derrick.
....	Simple Engine........	100	Machine Shop.
....	Simple Engine........	25	Pump Sta. Rock Crusher.
		180	

In this paper no attempt will be made to give more than a few facts in regard to the mining operations of the Homestake Mine. Much of the matter contained herein may sound commonplace to many experienced mining men. If any of the methods described are original, it should be borne in mind that they have stood the test of actual mining operations in a mine where a large daily tonnage at a minimum cost is the object sought.

Illustration No. 1 is a photograph of some of the company buildings and a part of the open cut at Lead. No. 2 is a view of the Ellison Hoisting engine. No. 3 shows the sand vats of Cyanide No. 1.

SHAFTS.

The Homestake Mining Co. is now operating six shafts, the Ellison, B. & M., Golden Star, Golden Prospect, Old Brig. and Golden Gate. Three of these, the Golden Gate, Old Brig and Golden Prospect, have reached the 800 foot level, while the Ellison, B. & M. and Golden Star have reached the 1,100 foot level.

Four of these shafts are located on the hanging wall side of the ledge, one on the foot wall side and one is sunk in the ledge. All are connected by permanent drifts driven in the "country" rock.

The Ellison is a large, three compartment shaft, two compartments being used for hoisting and one for ladders and pipes. The two hoisting compartments are 10x5 feet in the clear. The third or ladder compartment is 10x6 feet in the clear. The Golden Star and Old Brig each have three equal compartments, 4 feet, 6 inches by 5 feet in the clear. The B. & M. is similar, with the exception that there is a fourth compartment reaching the 300 foot level in which is operated a third cage. The Golden Prospect originally had only two compartments, one being 9 feet, 6 inches by 5 feet in the clear and the other being 4 inches wider to provide for pipes. Later a third small compartment was added from the 500 foot to the 800 foot levels, in which are placed the ladders and pipes. The timbering in the Golden Gate shaft is slightly different from the other standard three-compartment shafts but the dimensions are the same.

After the first 200 feet from the surface, the shafts, with the exception of the Golden Star, are sunk in hard carbonaceous wall slate. The dip of the slate intersects the axis of the shafts at angles varying from sixty to ninety degrees from the horizontal which makes blasting the cut quite difficult.

During the early history of the Homestake and Associated Mining Companies these shafts were not connected by underground workings. Now each is a factor in the mining operations of the company and through each is carried on some part of the work incident to mining about 4,000 tons of ore each twenty-four hours.

When sinking is resumed in any of these shafts, it is necessary to provide hoisting facilities to carry on this part of the work independent of the regular cages. This is accomplished by using a small twenty-five horse power hoisting engine driven by compressed air, which is placed on the lowest working level of the shaft. In the operation of this engine, it has been found necessary to re-heat the air before it is used in the cylinder and this is done by passing the air through several coils of pipe surrounding a small sheet iron stove, in which coke is used for fuel. A bucket with a capacity of one-half ton is used instead of a cage to hoist the rock and men. No guide head is used, plank being nailed on the inside of the timbers of the hoisting compartment to prevent the bucket striking on the timbers. A small sheave wheel is placed just above the station set.

The men are well protected by bulk heads thrown across the shaft just above the sheave wheel or below the station on which the engine is located. The shaft is sunk 200 or 300 feet in this manner; stations are opened on the new levels and cross cutting is commenced. Drifts are run to other shafts of the belt, which are tapped by small raises. These raises serve the double purpose of providing ventilation and acting as a chute through which the rock is dumped as the sides are stripped down to make room for the regular shaft timbers.

Miners working in the shaft are usually under contract. The engineers, landers to dump the rock from the bucket into the cars, and timbermen are paid by the day. The company furnishes tools and sharpens drills, the contractors furnish candles, powder, and pay for repairs on machines and assist in placing timbers.

The following condensed statement shows the number of men employed, wages paid, and the approximate cost of sinking at the Ellison shaft.

The number of men employed on each shift is:

4 Miners—Contractors.
1 Engineer.
1 Lander.
1 Timberman.
1 Carpenter, 2½ shifts framing one set of timbers which provides for six feet of the shaft.

The actual time consumed in placing one set of timbers is eight hours, but the timbermen are kept busy one shift in each twenty-hour hours running timber from the saw mill and putting in extra bracing.

The cost of a set of timbers in place for the Ellison shaft is:

Timber			$26.85
Framing Timbers			10.00
4 Miners, 1 shift	@	$4.00	16.00
1 Lander, 1 shift	@	3.00	3.00
1 Timberman, 1 shift	@	4.00	4.00
1 Engineer, 1 shift	@	4.00	4.00
Total			$63.85
Cost per foot			10.64

The average progress made is one foot in two shifts, including the placing of the timbers. The cost per foot of shaft, assuming that the contractors made $4.00 per shift, which is the usual price paid to shaft men, is as follows:

8 Miners	@	$4.00	$32.00
1 Timberman	@	4.00	4.00
2 Engineers	@	4.00	8.00
2 Landers	@	3.00	6.00
⅙ Set timbers, lumber and framing			10.64
Total			$60.64

The Rand machine is used for drilling, mounted on a bar. Electric exploders are used to fire the blasts in all shaft work, forty per cent. dynamite being used as in drifting.

Station sets are from twelve to sixteen feet high. Two or three extra sets are put in on the stations which serve as braces to the long sets of the shaft and support the roof of the station.

Formerly all station floors were covered with sheet iron. On all new levels, however, the track rails are laid to the edge of the shaft and in line with the cage rails. Two tracks are laid up to the shaft, one to each hoisting compartment. A diamond cross over, located close to the shaft, permits the loaded cars to be run on to either cage and the empties to be returned to the empty track. A third track passes around the shaft when the station is opened on both sides. The cages are made at the company's shop and are provided with the usual safety appliances. All new cages are made with the chair and operating level on the cage, so that the cage tender always has control of the chairs. Double decked cages are used at the Ellison shaft, which carry two cars on each deck, the Golden Prospect cage carries two cars on a single deck and all other shafts have small single decked cages. Skips have not been used in any shaft now belonging to the Homestake company.

DRIFTING.

Three dimensions are used for "dead work" drifts, depending upon the use for which they are intended; prospect drifts 6x7 feet, single track working drifts 7x8 feet, double track working drifts 12x8 feet.

Prospect drifts are usually driven under contract, two men are employed on each shift. Each man is a party to the contract. The miners car their own rock, furnish powder, fuse, candles, shovels, and picks and pay for the repairs on machines. The company furnishes machine and drills, lays all track and puts in air pipe as needed. The price paid is from $5.00 to $8.00 per running foot, depending on the character of the rock.

No timbering is required except in a few isolated places where a soft seam is encountered necessitating standing a few tunnel sets.

In running the 7x8 single track drifts either one or two machines are used for drilling mounted on a horizontal bar. The bar has two decided advantages over the post; namely, the machine may be set up as soon as the blasting is finished and drilling for the next round goes on while the shovelers are removing the rock, and again the sides of the drifts are carried much more evenly than with the post which makes it possible to lay a straight track and allow ample room for a ditch on one side.

Mounting the machine on a bar instead of a post has only lately been tried in the Homestake Mine and the result has proven very satisfactory. Machine men say that it is easier to move the machine from one position to another which fact is an advantage to the miner and will in the end materially advance the interest of the company.

The following tabulated statement shows the cost of five feet of average 7x8 feet drift.

1 Miner,	2 shifts, drilling............@ $3.50........		$7.00
1 Helper,	2 shifts, drilling............@ 3.00........		6.00
1 Miner,	1 shift, blasting............@ 3.50........		3.50
1 Helper,	1 shift, blasting............@ 3.00........		3.00
1 Shoveler,	2 shifts...................@ 3.00........		6.00
Explosives, powder, caps and fuse........................			10.75
Blacksmith labor, repairs of machine and machine shop labor.			4.85

Total cost of five feet of drift......................$41.15

Cost per foot.. 8.23
Cost per ton.. 1.47

The above cost statement does not include the cost of laying track or putting in air pipe. This was left out because track and air pipes are a part of the permanent improvements of the mine and not all of this item should be charged against any particular drift.

The holes are drilled about six feet deep and fully five feet of ground is blasted out with each round. Ingersoll Sergeant rock drills D–24 and D–32 are used almost exclusively in drifting. These drills seem to require less skill in manipulation, the repair account is smaller and they stand more hard knocks than any machine yet tried. Where two machines are used

mounted on one bar one helper, who receives the same wages as the miners, attends to both, and the same number of holes can be drilled in a little more than one-half the time required when only one machine is used.

The number of cubic feet of rock excavated at each round is (7x8x5 feet), 280 cubic feet. Ten cubic feet of this rock will weigh one ton, so that 280 cubic feet equals twenty-eight tons. Dividing $41.15 by twenty-eight gives $1.47 as the cost of excavating one ton of rock from the face of a 7x8 foot drift. It may be of interest to note in connection with the weight of the rock that the shovelers will have to remove from twenty-five to thirty cars of broken rock. Homestake Mine cars have a volume of twenty cubic feet. This would indicate that the volume increases 100 per cent.

Tabulating the cost of double track 12x8 foot drifts in a similar manner gives the following:

1 Miner,	3 shifts, drilling. . . . . @	$3.50	. . . .	$10.50
1 Helper,	3 shifts, drilling. . . . . @	3.00	. . . .	9.00
1 Miner,	1 shift, blasting. . . . . @	3.50	. . . .	3.50
1 Helper,	1 shift, blasting. . . . . @	3.00	. . . .	3.00
1 Shoveler,	3 shifts. . . . . @	3.00	. . . .	9.00
Explosives, powder, caps and fuse.				16.90
Blacksmith labor, repairs and machine shop labor.				6.50

Total cost of five feet of 12x8 foot drift. $58.40

Cost per foot. 11.68
Cost per ton of rock excavated. 1.22

This statement shows a decided reduction in cost per ton in comparison with the smaller drift, and this is as it should be, for it is well known that rock can be broken more easily from a large face than from a small one.

Cross cuts and drifts in ore are from eighteen feet to twenty-four feet wide and ten feet to twelve feet high. These being more or less irregular, no estimate of cost can be had. It is perhaps needless to say that per ton of rock excavated the cost is materially less than in smaller drifts.

As there is very little water in the Homestake Mine, considering the extent of the openings, the main working drifts are run on a nearly level grade. No working track has a grade greater than one per cent. and most of the long headers have a grade of two-tenths per cent. The drainage is provided for by a ditch one to three feet deep at one side of the drift. Long stringers of six-inch lagging extend across the ditch and are supported by lagging at one end while the other end rests on the ground. These stringers are spaced about twelve feet apart and serve as ties. Short ties are used between the stringers.

In some cases the air pipe is laid on the stringers, directly over the ditch. This method is preferable where there are not too many cross cuts, as it does away with a considerable expense in providing hangers for the pipe. When the pipe is suspended, the electric wires may be supported by wooden cross arms which are fastened to the pipe by a yoke made of round or strap iron.

When required, tunnel sets are put in, but they present no unusual features to the average mining man.

RAISES.

Raises are made for one of four purposes; namely, ventilation, ore storage bins, transferring ore or waste from one level to another or to provide an opening through which waste may be dumped into a finished stope.

Permanent raises for ventilation, ore bins, etc., are located in the country rock far enough from the ledge to be undisturbed by mining operations.

The purpose of the waste raise is to provide a storage for waste taken from dead work when not immediately needed in the stopes, and are so arranged that waste may be drawn out or dumped in on any level. The main waste raises are connected with the surface and the porphyry which here forms a cap overlying the vertical formation, is drawn through them and used for filling.

The method of procedure is as follows: On one or several levels, if convenient, cross cuts are driven to the point from which it is desired to make the raise. Care must be taken to have the cross cuts on the different levels alternate on each side of a vertical plane. When the cross cut has been driven far enough beyond the position of the raise to provide a passing track for empty cars the drift is widened, four sets of regular stope timbers are put in in the form of a square and an ordinary board chute located in one set. Above these timbers, the raise is gradually drawn in to a 6x6 foot raise. From this point the raise is carried up in the usual manner. Sprags across the raise about five feet apart are used to place the working platform on and serve as a ladder way. The smaller size machine drills are used for drilling and are raised and lowered by rope and pulley. The raise on the level below is located about fifteen feet on either side and is raised straight to the upper level. After connection has been made an inclined by-pass is made to connect with the upper raise above the timbers. Another small inclined raise connects the lower raise with the cross cut at the top of which are located the grizzlies. The bars forming the grizzly should be spaced about one foot in the clear so that no large rock can get into the raise. The successful operation of these continuous raises depends in a great measure upon the grizzlies. The by-pass is closed by a gate made of steel plate, sliding in cast iron grooves fastened to upright timbers, and operated by rack and pinion. This gate is used in the main ore bins under the crushers and is made at the company's shop. It has the advantage of being easily operated and can be operated gradually which prevents rushes of rock. Similar arrangements are made on the other levels. As soon as one raise is made it can be put into service and the raises from the lower levels finished when needed.

In making an ore storage bin, the timber sets are carried up six or seven posts high. When connection is made with the level above the timbers are removed with the exception of the sill floor in which the chutes are located. The by-pass and small raise to the grizzlies are located as in the waste raise.

Ingersoll A–32 machines are usually used in making raises. One miner and helper will make a 6x6 foot raise 100 feet high in ordinary ground in sixty shifts, provided they are not required to car the rock.

Blasts are fired by ordinary caps and fuse. Forty per cent. dynamite is used as elsewhere in the mine. The manner of placing the holes is similar to that used in small drifts.

When a stope has been worked nine or ten floors high, raises are put up in convenient places to the level above through which the filling is dumped. These raises are all in ore and consequently pay their own way.

When a raise is to be made near the face of a long tunnel where the air is bad some artificial means of ventilation must be provided. A device which I believe originated with one of the assistant foremen has proven very successful. A six or eight inch light iron pipe is laid from the entrance of the drift to the foot of the raise. Near the entrance a small one-half inch pipe is tapped into the main air pipe and brought down to and into the large pipe with the end which projects into the pipe turned out. It has been found that a very small amount of air having a pressure of from seventy-five to eighty pounds will effectually clear the raise in a few minutes. This device is used also in running long drifts. An exhaust fan would possibly be a more economical machine so far as power is concerned but the first cost would be greater and the air is used only when the machine drill is not in service.

STOPING.

During the year from June, 1902, to June, 1903, 1,279,000 tons of ore were milled by the Homestake Mining Co. Approximately eighty per cent. of this was mined on the different levels from the 100 foot to the 1,100 foot. The remainder came from open cuts and from "draw raises" by means of which, as will be explained later, the crushed ore left in the roof of old stopes is drawn out.

OPEN CUTS.

The ore obtained from the open cuts is broken down from the sides into openings that connect with the regular levels of the mine. These openings or raises are provided with chutes on the different levels from which the cars are loaded. The loaded cars are made up into trains of from four to eight cars, according to the grade of track and hauled to the shaft by horses.

The ore taken from open cuts is mined very cheaply. Two miners whose wages are $3.50 per day will break, on the average, 200 tons in one shift. Two men are employed at the chutes, one to break the rock at the grizzlies and one to load the car. The grizzlies are located two or three floors above the chute. If the haul is long two horses with two drivers are required to haul the ore to the shaft. The cost per ton of mining and delivering to the shaft is as follows:

2 Miners	@	$3.50	$7.00
1 Grizzly man	@	3.50	3.50
1 Chute Drawer	@	3.00	3.00
2 Drivers	@	3.00	6.00
2 Horses	@	.90	1.80
Blacksmith labor			.50
Explosives, .026 per ton			5.20
Total			$27.00
Cost per ton			.13½

Ore and waste is blasted down together and the waste is sorted at the chute and used for filling.

In one of the open cuts, now being worked, the process of mining is much like the work of making a very deep thorough cut through a hill some 300 feet high. Tracks are laid from the crushers to one end of the cut and the rock is blasted down, loaded into cars and trammed directly to the crushers.

ROBBING OLD WORKINGS.

The pillars and backs of old stopes which were left in place in the older workings of the mine are now being removed by a system of "draw raises." A description of the method of attack may be of interest to many who are face to face with the problem of robbing old workings in vertical formations.

It is very important to know the exact location of these pillars and roofs before starting this system. If the maps and stope records have been faithfully kept this becomes an easy matter; otherwise the memory of some old employe must be depended upon to give the necessary information.

Having located the ore, a "nine post raise" which consists of four regular stope sets arranged in a square, is put up either on the foot wall or on the hanging wall side; preferably the foot wall. This raise is carried up a sufficient height to reach the ore above the waste filling. Grizzlies are put in on the floor next to the top and the sides of the raise are carefully lagged to protect the men. The run of ore is then started by a blast or by barring. A man stationed at the grizzly breaks the rock so that it will pass through into the chute which is located on the sill floor.

The timber required to make a nine post raise four floors high, is thirty-six posts, twenty-four caps, twenty-four ties, about 150 lagging, four stope ladders and one ordinary board chute. The grizzlies are made by placing two or three twelve inch timbers, spaced twelve inches apart, over one set and protecting these by pieces of sheet iron curved to fit. The posts on the top floor may be protected in a similar manner, from the blasts and from running rock.

Sixteen thousand tons of ore were taken from one raise in six months; and this was done in a place where vain attempts had been made to reach the ore by carrying up timbered stopes through the old fill.

Should the ore be too solid to run, the raise will serve as a manway to a timbered stope started on top of the fill.

Whenever a run of waste is encountered it is drawn down and used for filling in other parts of the mine and the ore from upper levels will follow the waste down.

TERRA CAVING SYSTEM.

Stoping without timbers was carried on in the Deadwood-Terra Mine which is now a part of the consolidated Homestake, for a number of years prior to the time when the Homestake assumed control.

A cross cut from the shaft cuts the ledge which was opened from wall to wall throughout the entire length. All the ore was removed from the sill floor excavation. As soon as the ledge was sufficiently developed a drift was driven in the footwall approximately parallel and about twenty feet from the ore, and openings made from the drift into the ore chamber at convenient intervals. The ore was then broken down and the surplus removed through these openings. As the miners used the broken ore as a staging to work on, only about forty per cent. of the ore could be removed until the stope was finished.

The stopes worked by this method are from thirty to fifty feet wide and the dip of the ledge is seventy-five degrees.

Stoping may be carried on on several levels at the same time, provided sufficient back is left in between levels. When the level above is finished this back may be caved and all the rock removed on the level below.

No timber is required other than a few lagging for staging, but the method is not applicable to wide ledges.

SQUARE SET TIMBERING.

Until within the last two years all stopes in the Homestake and Highland Mines were timbered by the square set method. An enormous amount of timber was required to timber the excavations and the work of handling the timber through the shafts made it difficult to hoist sufficient ore to supply 900 stamps.

No unusual features are introduced in this method of timbering so that no detailed description will be given. The sets are six feet square, sill floor sets are nine feet high and all upper floor sets are eight feet, five and one-half inches. A sill floor set contains 324 cubic feet or about thirty-two tons, and an upper floor set 304.5 cubic feet or thirty tons.

All stope timbers are shipped in convenient lengths to the company's saw mill which is located at the Golden Prospect shaft. Here they are sawed and framed as needed. Most of the timber used is native, although some Oregon timber is used for caps and ties.

With labor at $3.00 per day, it has been found that the framing of what is termed a set of timber (a post, a cap and a tie), costs sixty cents and the cost for sawing lagging is five cents per running foot.

Two timbermen whose wages are $3.50 and $3.00, working one shift, will stand timbers, build chutes and do all necessary repairing in a stope which furnishes 100 tons of ore in twenty-four hours. This makes the cost of labor for standing the timbers six and one-half cents per ton.

All chutes are the ordinary board type, with bottoms made of lagging and sides of two inch plank. The rock is held back by two boards, one above but not directly over the other.

On the upper levels of the mine where the ledge was broken and comparatively narrow, the sill floor excavation was made from wall to wall and sometimes for the entire length. The sill floor timbers were then put in and the stopes worked where convenient. When the ledge began to widen this method proved disastrous and thousands of feet of lumber were used to make bulkheads in a vain endeavor to keep the stopes open.

A new method was inaugurated on the 600 foot level under the supervision of Mr. W. S. O'Brien, mine foreman, which is being used on all new levels. The plan is as follows: A cross cut is driven through the ledge from the central shaft which is called the main cross cut. The ore body is then developed by driving a twenty-four foot drift along the foot wall. From this drift rooms sixty feet wide are opened across the ledge with sixty-foot pillars between each stope. Beginning with the main cross cut a pillar is left on each side, then a sixty foot room, a sixty foot pillar and so on to the end of the ore body. These rooms and pillars are numbered north and south of the main cross cut. No. 3 Stope North would be the third stope north of the cross cut and No. 4 Stope South, the fourth stope south of the cross cut.

Some difficulty was experienced in keeping the sides of the rooms straight while the sill floor was being opened. To overcome this, sills were laid in the foot wall drift to lines given by the surveyor and the miners took their lines from these. Each stope contains eleven lines of sills. When the system once became established no difficulty was experienced in keeping the room of uniform width and the sides comparatively straight.

When a stope has been worked and filled the pillar may be attacked and by lacing the sides next the fill all the pillar may be removed.

HOMESTAKE SYSTEM.

However satisfactory this method of blocking out the ore body may be it did not solve the timber problem. After much deliberation and discussion it was decided to try stoping in these rooms without timbers. Again the practical ideas of the mine foreman and his assistants worked out a system which seems to be well adapted to the conditions existing in the Homestake Mine and which for the sake of convenience has been called the "Homestake System of Stoping."

The level is opened by the room or block method and sills are laid in the rooms the same as for timbered stopes. When the sills are in, three lines of track are laid running lengthwise of the stope but crossing the ledge with as many cross tracks connecting them as are necessary. The sill floor posts are put up and lagging placed over the top, the tracks are protected by double lagging on top and the rock is prevented from running in at the sides onto the tracks by lagging or slabs spiked to the posts.

As soon as the timber is in position the mining operation begins. The ore is broken down and allowed to fall through the lagging entirely filling the sill floor sets, with the exception of the carways. The lagging, which serves merely as a staging, is removed as fast as the sets are filled with broken ore. No rock is removed from the stope until this filling is finished. When the next cut or breast is carried across the stope some ore must be removed to make room for the miner.

In the large stopes two D-24 Ingersoll machines are employed with from one to two "baby" machines which are used to drill blockholes in the large boulders.

Somewhat of an innovation in machine drills for block holing has been introduced. A small pneumatic hammer, such as is used in shops for chipping and caulking, fitted with a rotating movement operates a small drill bit about one inch in diameter. This machine will drill holes from six inches to twelve inches deep and has so far proven very successful in block holing large boulders in the open stopes.

As there are no timbers to break, no limit is placed on the miner as to the amount of rock he may bring down at one blast. The stope should be finished as quickly as possible so that the broken rock may all be removed if needed. Consequently large slabs of ore are blasted down and these must be broken up to regular car size either on top of the pile or on the sill floor as it is drawn down by the shovelers.

On account of the uneven size of the rock chutes are not generally used in these stopes but the car men shovel the ore into cars from the level of the track, there being as many places to shovel from as there are spaces between posts along the track. However, where the rock is soft and where it breaks fine, chutes are used to advantage.

Should a large rock come down which the shoveler cannot break with a rock hammer, he moves his car to another opening until the "block holer" comes around.

Two or three regular sets on each side of the stope are carried up as fast as the stope is worked in which are placed the ladders and air pipes. These open sets also assist in ventilating the stope.

When the stope is worked up eighty or eighty-five feet, raises are made to the level above through which the filling is to be dumped, and the ore is then drawn out. While the ore is being drawn out, the walls and roof are carefully watched and all loose material is dressed down. No accident of a serious nature has occurred in one of these stopes during the two years in which this method has been employed that could in any way be attributed to the method.

When one end has been emptied of ore, a section of the sill floor is lagged and the filling is dumped in until it begins to run over the lagging. In this way the filling follows the shovelers and the walls of the stope are supported at one end by the ore and at the other by the waste.

When small ore bodies are worked by this method, no pillars are left in, but when one section is worked up a sufficient height, another section is started at one end and the ore is left in until the entire body is worked.

Stoping without timber is not confined to the Homestake Mine, but there are certain features of the method as employed here peculiar to the Homestake and which are considered necessary to suit our conditions. In the Treadwell Mine the sill floor is not opened on the station level but drifts are run in the ledge and raises put up from these drifts to a level some fifteen feet above.

No timber is required except for chutes but a back of ore is left in which takes the place of the sill floor timbers used in the Homestake. The cost of the timber would, in a great measure, be offset by the cost of making raises and putting in chutes. The accessibility of the ore is another advantage in favor of the Homestake method and becomes a necessity in a mine which furnishes nearly 4,000 tons of ore every twenty-four hours.

As only a small per cent of the ore can be removed before the stope is finished, there is of necessity a large reserve always on hand, which allows the mine to lay off whenever desirable. The present broken ore reserve in the mine is nearly one million tons.

Following is a comparative table giving the cost of timber in a stope which has been worked and timbered by square sets and the same stope if worked by the "Homestake Method."

TIMBERED STOPE.

Name of Piece	No. of Pieces	Lin. Ft. or Ft. B.M.	Cost of Material	Labor, Sawing, Framing	Total
Sill Floor Posts........	421	3,650	$474.50	$96.83	$571.33
Upper Floor Posts.....	2,077	16,616	2,160.08	477.71	2,637.79
Caps.................	2,410	13,255	1,723.15	506.10	2,229.25
Ties.................	2,261	12,435	1,616.55	474.81	2,091.36
Sills, 203 long, 382 short	.....	4,537	266.85	22.69	249.54
Lagging.............	13,020	75,906	3,795.30	379.53	4,174.83
Lagging strips........	2,410	4,025 B.M.	64.82	30.00	94.82
Wedges.............	2,352	784 B.M.	13.33	11.76	25.09
47 Sill Floor Chutes— Complete..........	.....	....	....	....	327.93
215 Upper Floor Bins— Complete..........	.....	....	....	....	824.12
Ladders..............	14	117 B.M.	1.99	3.50	5.49
Labor placing timbers& Chutes...........	.....	....	....	....	4,745.00
Breakage (10% of Lagging, 5% posts, caps ties).............	.....	....	....	....	793.97
			$11,174.74	$2,057.08	$18,770.52

STOPE WORKED BY "HOMESTAKE METHOD."

Name of Piece	No. of Pieces	Lin. Ft. or Ft. B. M.	Cost of Material	Labor, Sawing, Framing	Total
Sill Floor Posts.......	421	3,650	$474.50	$96.83	$571.33
Caps..............	410	2,250	293.15	86.10	379.25
Ties...............	381	2,095	272.35	80.01	352.36
Sills, long............	203	2,436	121.80	12.18	133.98
Sills, short...........	382	2,101	105.05	10.50	115.55
Lagging..............	1,752	10,214	510.70	51.07	561.77
Lagging to protect track	764	4,454	222.70	22.27	244.97
Relief Lagging........	1,684	13,472	673.60	67.36	740.96
Wedges.............	200	66 B.M.	1.12	1.00	2.12
MANWAYS.					
Upper Floor Posts.....	96	768	99.84	22.08	121.92
Caps..............	48	264	34.32	10.08	44.40
Ties...............	48	264	34.32	10.08	44.40
Lagging, Floors.......	96	560	28.00	2.80	30.80
Lagging, Sides........	720	4,197	209.85	20.98	230.83
Drift Pins...........	1,440	457 lbs.	22.85	....	22.85
Laders.............	28	235 B.M.	4.00	7.00	11.00
Labor Standing Sill Floor Timbers.......	....	....	....	....	758.16
			$3,108.15	$500.34	$4,366.65

73,000 tons taken from this stope.

18,770.52÷73,000....$0.257 per ton by former method.

4,366.65÷73,000....$0.060 per ton by Homestake method.

$0.257—$0.060....$0.197 saving per ton.

THE METALLURGY OF THE HOMESTAKE ORE.

By C. W. Merrill, B. S., Member Amer. Inst. Mining Engineers, Member
Inst. Mining and Metallurgy, Assoc. Chem. and Metallurgical
Soc., S. A., Lead, South Dakota.

[Revised from paper. Read before Black Hills Mining Men's Association,
February 19, 1903. Published by Permission of A. I. M. E.]

I. THE PROPERTY.

The Homestake Mining Co. has acquired through consolidation the ground and equipment of the Father De Smet Consolidated Gold-Mining Co., the Deadwood-Terra Mining Co., the Caledonia Gold-Mining Co. and the Highland Mining Co., at and near Lead, Lawrence County, South Dakota, in the northern portion of the Black Hills. The company owns or controls 250 claims, comprising 2,616 acres, and covering about 8,000 feet along the strike of the lode.

At the surface there are several veins, of which three have united in depth, where the main vein ranges from 300 to 500 feet in width. The 1,100-foot level is the lowest at present. The rock of both walls is, so far as known at present, carbonaceous slate, and the country is penetrated by a system of porphyry dikes, and in some places capped with porphyry.

The output of the company up to January, 1903, has been, approximately, $70,000,000.

II. THE ORE.

The oxidized, open-cut ore is nearly all treated in the three mills on the northern part of the property, which are as follows: The Mineral Point (formerly the De Smet), of 100 stamps, the Monroe (formerly Caledonia), of 100 stamps, and the Pocahontas (formerly the Deadwood-Terra), of 160 stamps. A cyanide plant, to treat the leachable portion of the tailings from these North End mills, has recently been installed and put into operation at Gayville, or Blacktail as it is now known.

The Homestake lower-level ore, which comprises the greater part of that being milled at Lead, may be described as a hornblende, garnetiferous schist or slate, which has been crushed and infiltrated with free silica and pyrites, the latter being about seven or eight per cent. of the ore, and comprising pyrite, pyrrhotite and traces only of chalcopyrite and arsenopyrite.*

---

*The standards used in this discussion are the U. S. gold dollar; the U. S. short ton of 2,000 pounds avoird., and the value of an ounce of fine gold, $20.67. Percentages are given by weight, and not by volume. Sizings are classed as *coarse* (that portion of the sample which will remain on a 100-mesh screen; diameter of wire, 0.00433 inch, size of opening, 0.00575 square inch); *middles* (the material finer than the opening of the above 100-mesh screen, and coarser than the opening of the commercial 200-mesh screen as given below); and *fines* (the material which will pass such a 200-mesh screen; diameter of wire, 0.00216 inch; size of opening, 0.00312 square inch). A further subdivision of fines is also mentioned, which is based upon the granular, or angular, condition of one portion and the flocculent, or amorphous, condition of the remaining portion of these fines.

III. MILLING.

The ore receives its first crushing in rotary breakers at the hoists, and this product varies in size from that of sea-sand up to rock having an extreme dimension of four inches.   From bins at the hoists at Lead, the broken ore is trammed to the three mills, *i. e.*, the Homestake and Golden Star, containing 200 stamps each, and the Amicus (formerly the Highland), in which there are 140* heads, making a total of 540.

From the mill-bins the ore passes to the mortar, which is of the now celebrated Homestake narrow pattern, where it is crushed between cast-iron shoes and dies, the weight of the stamp when equipped with new iron being 900 pounds, the drop ten and one-half inches and falling eighty-eight times per minute.

The screen is of the steel-needle slot-type No. 8, and the bottom of the screen-opening averages ten inches above the top of the dies.

The long drop, high discharge and small area of screen openings produce an extremely fine pulp, about eighty per cent. passing a 100-mesh screen, and it is to the writer a most remarkable fact that under these conditions such a high stamp-duty is maintained, it being fully four tons per stamp per twenty-four hours.

This duty is possible only because: first, of the very favorable nature of the ore, the slate and pyrite crushing readily and the quartz being an excellent medium of attrition; secondly, of the large proportion of water used, being from eight to ten times the weight of ore crushed; and, thirdly, of the narrow mortar, which is only twelve inches wide at the lip.

This very fine and thin pulp is in the most excellent condition for amalgamating, which process is conducted both inside the mortar and outside, on four full-size plates in series (each 54x144x$\frac{1}{4}$ inch) to each mortar.   The first of these is a copper-plate, and the other three are silver-plated copper, the weight of plating being two ounces per square foot, and all silver-plating being done at the works.   The addition of the three silver-plates to each stamp-battery by Mr. Grier has proved one of the most valuable steps in the treatment of this ore, and has brought about an additional profit amounting to, approximately, $250,000 during the year 1902, over and above what would have been realized from amalgamation had the outside plate-surface been only that of the one copper-plate—which, by the way, is considered ample in many of the large modern plants of the day.

In connection with amalgamation, the practice at the Homestake conforms, as far as conditions will permit, to the theory that the maximum results are obtained when the temperature of the water used in the batteries is low enough to exert the minimum influence on the minerals of the ore; and it is contended that the plate-yield proves the correctness of this theory.

It would be interesting to investigate the question of amalgamation and finer crushing in other gold-producing sections, particularly in South

*This has now (Aug. 15th, 1904,) been increased to 240 stamps.

Africa, where the yield from this source is reported to be from fifty-five to sixty per cent., as compared with seventy to seventy-five per cent. at the Homestake. Perhaps finer crushing would not only greatly increase their amalgam yield, but also reduce the values lost in their cyanide-residues. This seems the more likely for that country, because their slimes have been proved to have value sufficient for secondary treatment, whereas this has not yet been proved at the Homestake, where the advisability of sliming such a large proportion of the ore has been a debatable point, because the slimes here contain only $0.85 to $1.10 in value per ton. But of this more will be said later.

The total cost of milling in the 200-stamp mills at Lead is, approximately 40 cents per ton.

CLASSIFICATION.

We now have a pulp containing eight or ten parts of water to one of ore; and much of the latter is so infinitesimally fine as to cause a visitor, who had watched an attempt to filter the slimes on a large scale, to say that, for an exemplification of the size of a molecule, he would advise the study of Homestake slimes.

The tailings as they leave the mill are sized, with the following result:

Coarse (remaining on 100-mesh), twenty-two per cent.

Middles (between 100 and 200-mesh), eighteen per cent.

Fines (passing a 200-mesh screen), sixty per cent.

That is, sixty per cent. of the particles issuing from the mortar have less than 0.00001 square inch of cross-section.

When the erection of the cyanide-plant had been determined upon, the question of a tailings-wheel to elevate the pulp and permit the location of the plant nearer the mills being under discussion, it was calculated that to elevate the tailings at a cost of about 2 cents per ton would cost the company, approximately, $140,000, on the proportion of the material then blocked out in the mine which would be available for leaching. In other words, for every cent per ton which could be saved in the secondary treatment of the leachable material, the company would profit ultimately to the extent of at least $70,000. Consequently, the plant was located, as shown in Fig. 1, about a quarter of a mile below the Lead mills; and the problems of transportation and of such classification as would permit the pumping-plant to return its former percentage of water to the mills, presented themselves. Another set of 16 settling cones 10 feet in diameter has recently been installed. The latter has been met by the installation of the upper cone-house, where twelve gravity-settling cones, seven feet in diameter and with fifty degree sides, throw off about half the water and, perhaps, one-fifth the solid matter, which latter is the very finest slime, of the following sizing, during 1902: *Coarse*, 0; *Middles*, 1.76; *Fines*, 98.24 per cent. The thickened slimes are, subsequently, settled out of this pulp, and a part of the water is returned to the mills.

From the bottom of the cones is drawn the thickened pulp, containing all of the leachable material and some of the slimes. This portion is transported by means of a twelve inch cast-iron flanged pipe on a minimum grade of 2.5 per cent., and with as few turns as possible, to the cyanide plant.

U. S. HOSPITAL, HOT SPRINGS, S. D.

The second step in the classification is carried out in the plant proper by means of six more gravity-settling cones, the overflow from which, of a like composition to that of the first twelve cones, is conducted to a collecting-tank, whence it is drawn for the purpose of sluicing out the leachable material after its treatment has been completed. The average sizing of this second settling-cone overflow for 1902 was: .*Coarse*, 0; *Middles*, 1.38; *Fines*, 98.62 per cent.

The under-flow from the second set of gravity-settling cones, which is now quite thick, passes to twenty-four (this number has since been increased to 36) sizing or hydraulic classifying-cones, which carry a device for discharging the sand and introducing the water, patented by the writer. By its means the admission of water does not result in currents of varying velocity, which latter always interferes with uniform separation of slimes from granular material.

These sizing-cones complete the classification, which has been a difficult problem, first, because of the extreme fineness of the pulp, and, secondly, because the writer was determined to avoid double treatment, which entails a largely increased installation and operating-coat, but which is necessary, unless a product be obtained practically free from slime.

The slime-overflow from hydraulic classifiers, had the following sizing average for 1902: *Coarse*, 0; *Middles*, 1.46; *Fines*, 98.54 per cent. As regards all slimes referred to, they will practically pass the 200-mesh screen, the middles being largely wood-pulp.

In fact, there is little doubt but that the importance of the most perfect classification possible will be recognized shortly as a vital consideration in the cyaniding of wet, crushed ore; and metallurgists will not follow the old German practice of *spitzkasten* and *spitzlutten*, which are very imperfect machines as compared with a cone-classifier or sizer for separating granular from flocculent material. The writer's judgment is that a scientific classification-system, by which all the granular or angular material may go to the leaching vats, and all the amorphous portion to the slime-plant, will in the future be a feature in designing a plant on which the greatest care and experimentation will be put, and the highest grade of technical skill utilized.

CYANIDE-TREATMENT.

By these three steps in the classification. we have separated the pulp into non-leachable slimes, comprising about thirty per cent. of the ore crushed, and practically all passing a .200-mesh screen, and a direct-leachable product amounting to, approximately, seventy per cent. of the tailings, which, although very clean and free from mud, is still of a very fine texture,—as the following sizing test, the average for the year 1902, will show:

Coarse, remaining on 100-mesh, 40.5 per cent.
Middles, 100 to 200-mesh, 30.8 per cent.
Fines, passing 200-mesh, 28.7 per cent.

While this fineness is notable, we find that, as the proportion of lower-level ore increases, we can treat an even finer product. A recent charge, containing as high as forty per cent. *fines*, maintained our normal

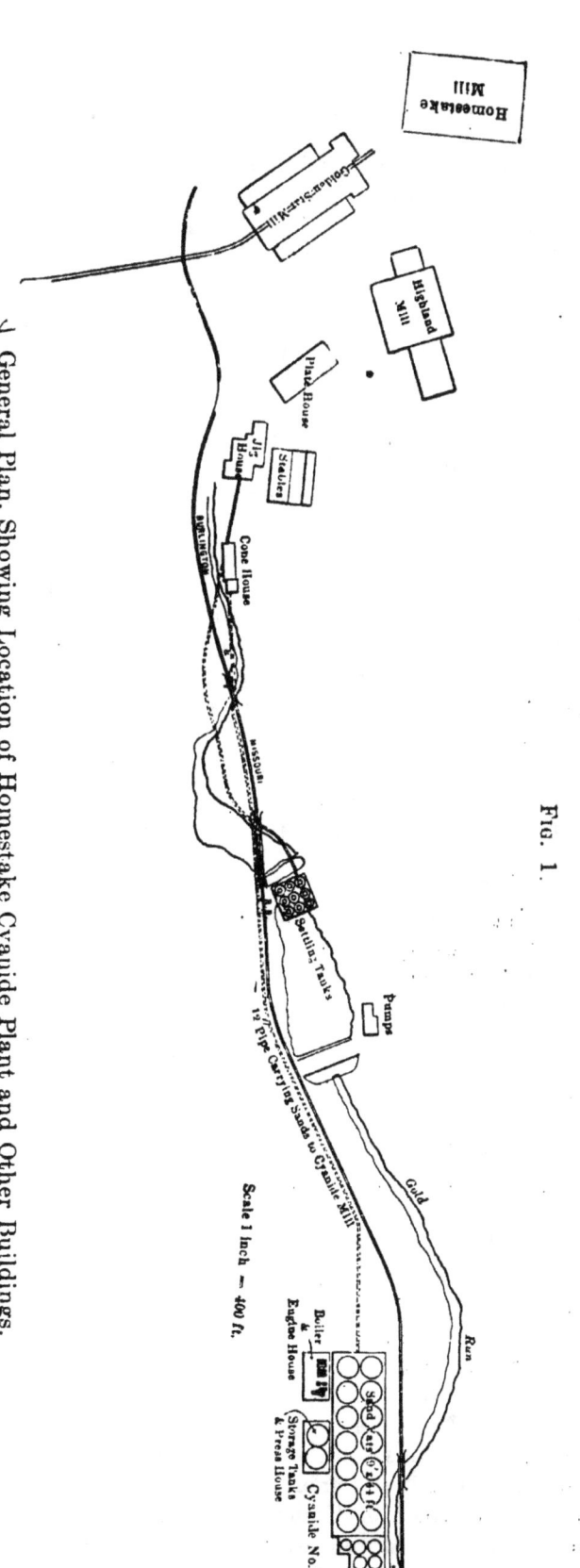

Fig. 1.

⌐ General Plan, Showing Location of Homestake Cyanide Plant and Other Buildings.

leaching-rate of three to four inches per hour throughout the treatment. This is undoubtedly due to the fact that the *fines* from the lower-level rock contain a greater proportion of angular or granular, and a smaller proportion of amorphous, hydrated of flocculent material.

The leachable pulp, which contains ten to twelve per cent. of pyrite. is now ready to go to the vats; and on the way lime is added in quantities varying from three to five pounds per ton. At first we tried adding this lime in the mills, as is done in Africa, but found that the amalgamation was most seriously affected thereby; not only was the plate completely coated, weeks being required to get it back in proper shape, but the tailings-values were largely augmented. This result only emphasizes the fact that the process must fit the ore, and that attempts to make an ore fit a process are useless. This practice of adding lime to the battery is, according to the writer's information, unanimously pronounced to work the best results in Africa, and to reduce the values in the slimes lost from amalgamation to half of what they are when no lime is used in the battery. In our case, however, we have demonstrated that the best results follow from crushing the lime wet into a running pulp which joins that from the sizing-cones, whereby there is less slacking and less loss of flocculent lime in the vat overflow, *i. e.*, in the water which overflows the vat, the sand having settled out. Not only is it of distinct advantage to have our lime go into the tank in un-slacked granules, but recent investigations are proving that the average size of these granules has an important bearing on the subsequent cyanide-decomposition and gold-extraction. This seems to be due to the fact that a low alkalinity, but one approximately constant throughout the leaching, is an important desideratum with the Homestake ore, on account of its considerable content of easily-decomposed sulphides. We are not, as yet, pre-pared to say what is the very best mesh-screen to use on our lime stamp-bat-tery, but at present we are using a wire-screen, the opening of which is eleven-sixty-fourths square inches. In this connection it should be said that only the purest lime should be used, the magnesia in the ordinary domestic limestone being objectionable for several reasons.

The classified pulp and the lime having comingled, the mixture passes to the distributor, which is of the garden-sprinkler, or Butters and Mein type.

There are two distributors, one for each row of vats, hung from a car-riage, which travels on a track, and the step of which rests on the top of the center-bottom discharge-gate of each vat, when the distributor is in operation. There are fourteen vats, each forty-four feet in diameter, nine feet deep in-side and holding 610 tons of sand. To fill one of these requires from eleven to eleven and one-half hours, which, with our equipment, permits of about five days' contact with solution, before it is necessary to recharge the vat. After filling, the drain-valve is opened, the top leveled, and the stronger of the two stock-solutions, of a strength of 0.14 of one per cent. KCN, is run on. The contact with this solution, including frequent drainages for the purpose of drawing in air, is maintained for about three days. The air-contact is very important in Homestake ores, owing to the presence of pyrrhotite or subsulphide of iron, which absorbs oxygen with great avidity, and which would greatly retard the dissolving action of the cyanide-solution were not large quantities of the essential oxygen introduced. The effluent solution during this period, having normally a strength of 0.10 of one per cent. of cyanide, is run to the weak precipitation tanks, of which there are two, each twenty-six feet in diameter by nineteen feet deep, and holding 300 tons of solution.

After the three days' contact with strong solution, the weak solution, normally of a strength of 0.10 per cent. KCN, is brought into the charge, and this contact is maintained for the remaining two days. The effluent solution from the charge during this period is run to the strong precipitation or rather collecting tanks, which are of the same size and number as the weak precipitation tanks.

Fig. 2 shows the interior arrangement of the works.

After contact with the weak solution has been completed, wash-water is brought into the charge, and the washing continued until the effluent solution is down to 0.03 or 0.02 of one per cent. in KCN and from 5 to 7 cents per ton in value.

The charge is now ready for sluicing, which operation is accomplished by two men, with three-inch hose, in about four hours, using the slime-water from the overflow of the second settling-cones. The four side-gates and one center-gate afford ample facilities for the discharging. The last inch or so of the sand is sluiced with clear water under seventy-five pounds pressure through one and one-half inch hose; and the eight ounce duck filter, under which is another of cocoa-matting, is washed clean. The vat is then field with water, and is ready for the next charging.

PRECIPITATION.

As stated above, the effluent solution resulting from the leaching with strong solution is run to the weak precipitation-tanks, and has a value of, approximately, $2.00 per ton and a strength of 0.10 per cent. KCN. When one of these weak precipitation-tanks is full, the stream is turned to the other, and the former is then ready for precipitation. It contains 300 tons of solution, which is brought into agitation by means of compressed air, and about sixty pounds of zinc-powder, in the form of an emulsion, is sprayed in during the agitation. The pump, which is of the compound, duplex, outside-packed, plunger type, is then started, and the mixture pumped through two large filter presses, thirty-six inches square, of the flush-plate and distance frame pattern, containing twenty-four frames, each four inches in depth.

While the gold, silver and excess of zinc remain in the frame and on the cloth, the barren solution passes through the cloth and on to the weak solution storage-tank below (of the same size as the sand-vats), whence it passes again to the sand as weak solution. Its value has been reduced by this operation from $2.00 to 5 or 10 cents per ton, being a precipitation of 95 to 97.5 per cent. The efficiency of this method lies largely in the fact that the cloths of the presses are coated with about one-eighth inch of powdered zinc and precipitate, so that every particle of solution, having to pass through the cloths, gets a molecular contact with the fine zinc, which is true of no other precipitation-process. The presses are run without opening for a month, at the end of which the press gauges indicate about ten pounds pressure, notwithstanding the fact that they then contain about a ton of precipitate worth, say, $50,000, when they are cleaned up by two men in about six hours, including the putting together with new cloths. Figures covering the labor of cleaning up $50,000 from zinc boxes and from electrolytic precipitation would form an interesting comparison.

We will now return to the effluent solution, resulting from the contact of the tailings with weak solution during the latter part of the leaching. This is run to the strong-solution collecting vats. When these are filled, they are strengthened to 0.14 per cent. KCN and pumped directly, without precipitation, to the strong-solution storage tank, of the same capacity as the weak storage, whence it goes on to the early treatment of the charge, as before mentioned. Its value is from 30 to 50 cents per ton. It will thus be seen that the strong solution of one day becomes the weak solution of the next day, and that the values are all accumulated in the weak precipitation tanks. The strong solution thus has an approximately constant value, that is to say, only one-half of the total effluent solution is precipitated, the other half being of a constant low value.

REFINING PRECIPITATES.

As the refining of cyanide precipitates is of some importance, owing to the well recognized losses taking place in the ordinary methods, which are from two to six per cent., a description of the process we use at the Homestake, in which the loss is less than 0.1 per cent., may be of interest.

The precipitate after removal from the presses, is treated first with dilute hydrochloric acid in a lead-lined mixing tank, equipped with a mechanical agitator, a hood, and a powerful exhaust-fan. After agitation, and settling, the supernatant liquid is forced through a filter-press by air pressure. Sulphuric acid is then added, agitation begun, and the mixture heated. It is then settled, and the supernatant solution put through the press, as in the case of the hydrochloric acid. Wash-water is then added to the mixing tank and the whole mixture put into the press, where it is further washed. The aggregate value of the acid-liquors and wash-water flowing from the press is less than $20.00 from $50,000 worth of precipitate. A portion of this value is recovered from a large settling tank, into which the effluent solutions flow, and the remainder constitutes the only loss we have been able to find in this process of refining.

The resultant, acid-treated precipitate is then removed to a large steam dryer, where a part of the moisture is expelled, but never all, and the precipitate mixed with litharge, borax, silica and powdered coke. When thoroughly mixed, it is sprinkled with a solution of lead-acetate and the whole mass briquetted under a pressure of 4,000 to 6,000 pounds per square inch. The zinc having been removed, and the briquettes having been dried, a borax-slag develops upon the outer surface upon being charged to the cupel, and they fuse quietly, quickly and at a low heat, without dust or volatili-

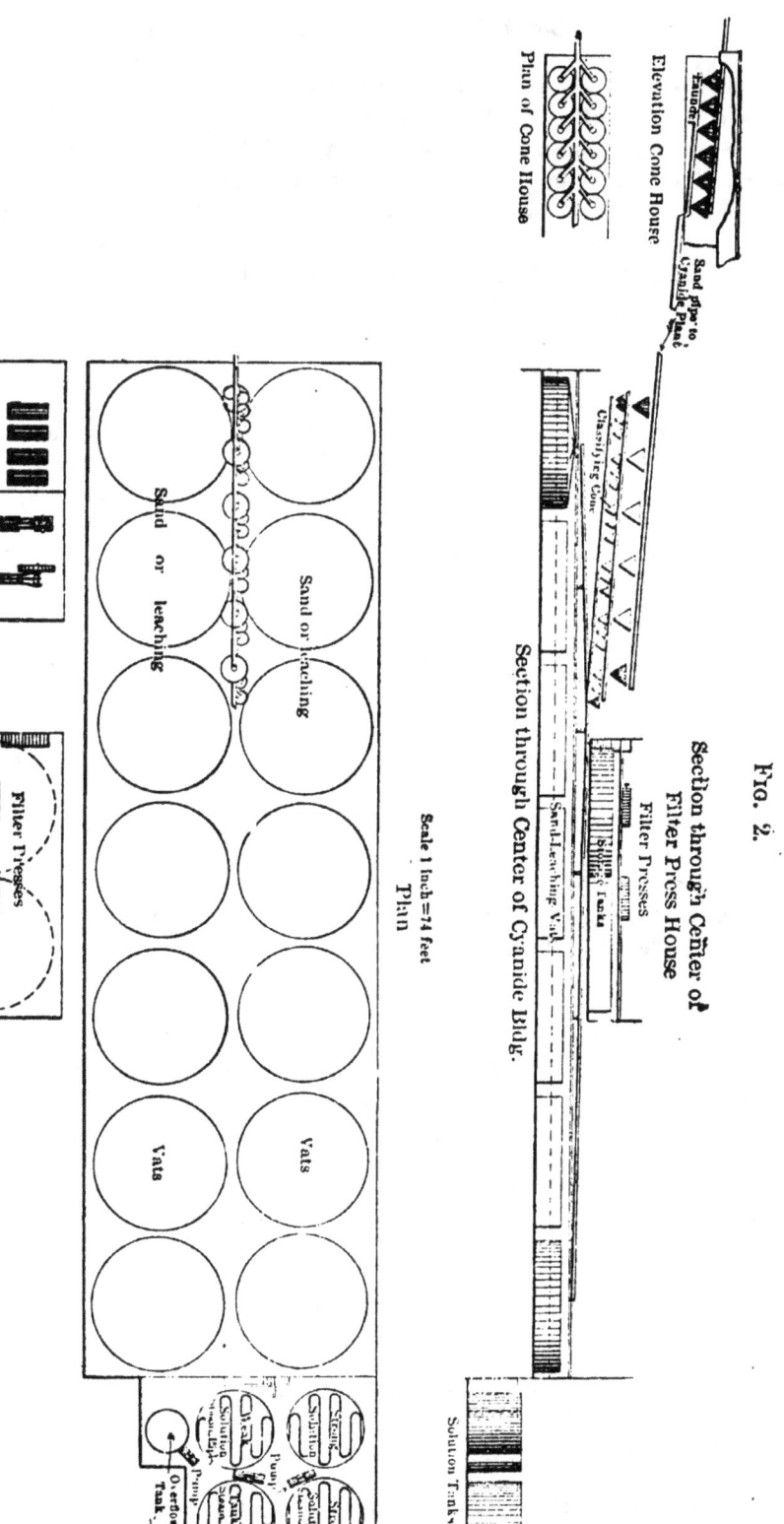

Fig. 2.

Plan of Cone House

Elevation Cone House

Sand pipe to
Cyanide Plant

Section through Center of
Filter Press House

Filter Presses

Classifying Cone

Sand-leaching Vat

Section through Center of Cyanide Bldg.

Solution Tanks

Sand or leaching

Sand or leaching

Vats

Vats

Scale 1 inch = 14 feet

Plan

Engine & Boiler House

Filter Presses

Plan of Filter Press Bldg.

Showing Storage Tanks.

Plans and Sections of Homestake Cyanide-Works.

zation losses. The lead absorbs the values, sinking to the bottom, and the slag is tapped off. All the slag having been removed, the lead is cupelled off as litharge, and the resultant metal, 975 to 985 fine, is ready to run into bars. The cupel-slag and the cupel-bottom are then put through the blast furnace, the lead content of the slag reduces to lead, which absorbs the values, and is drawn from the lead well in the usual manner. This lead is returned to the cupel at the next clean-up, the litharge from the cupellation goes to the next precipitate, and the blast furnace slag is worth less than $5.00 per ton.*

The total cost of this refining amounts to less than three-fourths of one per cent.; so that the Homestake Company realizes $20.52 per ounce for its cyanide gold, less the usual U. S. Assay Office charges on doré bullion, and the expressage to New York. These charges amount to between ten and eleven cents; and the net realization per ounce of fine gold precipitated is thus $20.42 in New York exchange. A parting-plant is now contemplated, which will make a further saving in this connection and enable the company to turn out fine gold and fine silver.

TONNAGE, PERCENTAGE AND COSTS.

TONNAGE.

The maximum monthly tonnage of this plant—which is ascertained by placing cubic foot boxes in many parts of various vats, determining the dry weights per cubic foot of sand and averaging a large number of such determinations,—was attained in October, 1902, when 40,236 tons, or 1,298 tons per day, were treated.† This gives to the Homestake Company the largest sand-treatment cyanide plant in the world; the next largest being, to the best of my knowledge, that of Simmer & Jack in South Africa.

PERCENTAGE.

As a comparison of the various assay-determinations and valuations with the bullion produced is always of interest, the following figures for the last half of the year 1902 are given:

EXTRACTION.

The extraction, as shown by the difference between charge and residue-assay multiplied by the tonnage, was $292,579.

PRECIPITATION.

The precipitation, as shown by the difference between assays of unprecipitated and precipitated solutions multiplied by the solution tonnage, was $301,233.

---

*The writer has applied for patents covering this process, which was first carried out experimentally during the latter part of the year 1900.

†Since this paper was written, another step in the classifying of the pulp has been added, with the result that this plant is now treating approximately 1,450 tons per twenty four hours.

GOLD IN PRECIPITATES.

- The amount of gold in precipitates, that is, the assay value of the precipitate sampled upon removal from the presses, was $302,895; the gold value of bullion shipped, $307,635, and the silver value, $2,874.

The average percentage recovered in bullion by the treatment for these six months is 74.7 per cent.

This is not as high a percentage of bullion as should be recovered from a porous or oxidized ore, or one in which the values are along cleavage-planes; but, in view of the facts that such a high percentage is recovered by amalgamation, that the values are very finely disseminated in the Homestake ore, and that the tailings are very low-grade, we feel, and all our tests so far have verified our conclusions, that it is the economic percentage, yielding the maximum net profit.

Many tests and experimental runs of the plant, looking toward a greater net yield, have been made, covering longer treatment, stronger and weaker solutions, extra oxidation with sodium and barium dioxide, and other similar reagents, varying alkalinities and alkaline reagents, etc. The question of separate treatment of concentrates and coarse sands has also been investigated, all with negative results. The conclusion of the writer in regard to this latter point is that, even if a higher net yield could be realized by separate treatment, which is contrary to the results of all our tests, a much greater proportion of the *fines* (passing 200-mesh screen) would have to be thrown off and wasted, entailing a serious net loss.

COSTS.

As to operating costs at the Lead cyanide plant, the following are the averages per ton for the year 1902, during which the average value of the material treated was $1.65 per ton:

Classification—Labor and Supplies............		$0.017
Treatment:		
Cyanide................................	$0.152	
Labor..................................	0.030	
Lime...................................	0.022	
Supplies................................	0.005	
	$0.209	0.209
Precipitation—Labor and Supplies............		0.026
Power—Labor and Supplies..................		0.051
Water......................................		0.026
Assaying—Labor and Supplies................		0.013
Refining—Labor and Supplies................		0.006
Miscellaneous..............................		0.005
Total.................................		$0.353

As compared with the above, the lowest costs I have seen authoritatively stated for other plants are as follows:

City and Surbuban, South Africa......................	$0.55
Geldenhuis Estate, South Africa......................	0.605
Geldenhuis Deep, South Africa.......................	0.62
Robinson, South Africa..............................	0.62
Worcester, South Africa.............................	0.72

The African costs refer, of course, to operations before the late **war** between England and the South African Republic; but they are the **only** figures available to me, and I do not think they have been reduced materially since.

As regards the Homestake slimes, which are not at present being treated, their assay-value ranges from $0.80 to $1.10 per ton, which is very much lower than that of any slimes now being cyanided elsewhere, and which does not offer much inducement to undertake their hydrometallurgical treatment from them, with a suitable plant.

THE CYANIDATION OF THE SILICEOUS ORES OF THE BLACK HILLS OF SOUTH DAKOTA.

By Charles H. Fulton, South Dakota School of Mines.

[Read before the Black Hills Mining Men's Association.]

The successful application of the cyanide process to the siliceous ores of the Black Hills contributes more than any other one factor to the present prosperity of the district. During 1903 about $2,200,000 was produced from the siliceous ores by cyanidation. While the cyanide process was introduced into the Black Hills about 1892, at the Rossiter cyanide plant, by the Black Hills Gold and Silver Extraction Company, the process was not actually demonstrated a great success until about 1900. Since then a considerable number of large plants have been erected which are operated successfully. At present sixteen cyanide plants are in operation in the Black Hills. In 1896 the Rossiter plant treated 6,500 tons of oxidized siliceous ores by cyanidation, while the Golden Reward and Keldonan Chlorination Mills treated 75,000 tons, so that the prospects for the cyanide process were not very bright. At the present time, however, no chlorination plants are in operation, and during 1903 about 550,000 tons of siliceous ores were treated by cyanidation. The application of the process has made available enormous low-grade ore reserves, which range in value between $3.00 and $12.00 per ton. Ore above $12.00 is generally considered of smelting grade, at the present time. The siliceous Potsdam ores are locally divided into two general classes: First, red ores (or the oxidized ores), and second, blue ores (or unoxidized ores), of which more will be said later. In general the red ores readily yield a fair extraction to the cyanide process, ranging from eighty to ninety per cent. in the district, the blue ores, however, fail in general to yield more than thirty to forty per cent. of their values to the ordinary process as at present applied. Some mines have large ore reserves of this blue ore which is below smelting grade, and it is but a question of time for the successful cyanidation of the "blue ores."

THE NATURE OF THE ORES.

The ores known locally as the "siliceous ores" are siliceous replacements of the lower layers of lime shales which rest on the Potsdam quartzite formation. However, the siliceous ores of the same general character occur also at higher horizons in the Cambrian, also in vertical veins, known as "verticals" and also as siliceous replacements of limestone in the carboniferous, as, for instance, the "Ragged Top" ores.

The ores consist in the main of silica in the form of quartz, from seventy-five to ninety per cent. This will range lower in the unaltered or "blue ores," than in the oxidized or "red ores." The blue ores contain from six to eight per cent. of fine grained evenly distributed pyrite, on the average, although this will run up to fifteen to twenty per cent. in some cases. In the "red ores" this pyrite has been altered to iron oxides which give the ore its characteristic red color.   Aside from these constituents the ores contain small percentages of lime, alumina and alkalies.

The presence of tellurium has been detected in some of the ores, pointing to the probable existence of tellurides of gold and silver.   Arsenic is found in small quantities in the ores, and antimony in the form of stibnite in small quantities is not infrequent.   Copper occurs in traces only.

The gold in the ores is present in but very small amounts in the free state, and that little is so rusty as not to be amalgamable.   The siliceous ores are in general very hard, but while the ores may be dense in some districts in others they are fairly open and porous, so that from the physical nature of the ores two types of mills have been developed.   First, the coarse, dry crushing plants, and, second, the fine, wet crushing plants. There is another type of plant, the fine, dry crushing plant, which method is in competition with the fine, wet crushing methods.

A typical plant of the first type (coarse dry crushing) is that of the

WASP No. 2 ON YELLOW CREEK, NEAR KIRK, SOUTH DAKOTA.

This mill was built during the early part of 1900, beginning operations about September.   It has been running continuously since then, and represents a modern plant, although improvements have been made recently, in increasing the capacity and in the method of filling the leaching vats.   It is a dry crushing mill, and has facilities for drying the ore.   The mine is within 250 yards of the mill, the ore being transported in trains of four cars, by mule haulage.   The cars hold 1,500 pounds of ore, are of the end dump type, and discharge into the main storage bin, of small capacity, which feeds to the crusher floor.

Nature of the Ore Treated.

The ore is of the quartzite variety from the Potsdam.   In the greater part of the ore the gold is carried seemingly on the cleavage planes of the quartzite, and is not finely disseminated as is usually the case, although the mine also has some of this latter class of ore.   The fact that the gold is carried in the cleavage planes makes coarse crushing permissible, manifestly a great advantage.   Very little pyrite is present, but some antimony in the form of stibnite is found, although only in small quantities, not sufficient to cause trouble in cyaniding.   The value of the ore varies considerably, generally running between $4.00 and $20.00 per ton, the greater part being nearer the first figure than the last.   The ore is at times quite acid, requiring considerable lime to neutralize it.

The Crushing of the Ore.

The mill site is very favorably situated as regards slope, full advantage being taken of gravity, there being but one elevator in the mill.   The capacity of the mill as built, was from fifty to sixty tons per day, but the installation of a new large tank raised this to 105 tons per day.

The Crushing and Screening Machinery.

The grizzley is eight feet long, four feet wide, with one and one-half inch spaces. It is inclined at forty-five degrees.

Coarse crushing is done in a No. 3 Gates crusher, breaking to a one and one-fourth inch ring. The capacity of the old No. 2 crusher was about sixty to sixty-five tons per twenty-four hours, and it was found necessary to put in a No. 3 Gates in order to come up to the increased tankage capacity when the new tank was installed. It has been found that a No. 2 Gates has a too limited capacity, even for a sixty ton mill, as the wet ore rapidly decreases its capacity. It is also the experience that the mouth openings are rather too small for the mine rock, so that too much sledging has to be done.

Coarse Rolls (1 set).—These are Gates rolls, twenty-four inches in diameter and fourteen inch face, making eighty revolutions per minute. The ore is fed to the rolls by Tulloch automatic feeds. The rolls are not housed, as the ore is damp enough to avoid much dust.

Finishing Rolls (1 set).—These are of the same make as above and the same size, but make 100 revolutions per minute. A Tulloch feed also supplies these rolls.

Stationary Inclined Screen.—This is seven feet long, built in two partitions, each one foot wide, giving fourteen square feet of screening surface. It is inclined at forty-five degrees. It is two mesh wire cloth, or four openings per square inch.

Combined Shaking and Stationary Inclined Screen.—This is the finishing screen. The first eight feet is a shaking screen and the balance, about eight feet, being stationary. It is two feet wide, giving a screening surface of thirty-two square feet. It is two and one-half mesh. This inclined screen is on the principle of the Berthelet separator, the material passing through it being considerably finer than the mesh would indicate. Practically, the product passing through this screen is not above six mesh. This is the size of the material cyanided.

The Leaching Department of the Mill.

There are four leaching vats, sixteen feet in diameter and seven feet deep. The vats are of Oregon fir, two and one-half inch staves, bound by seven five-eighths inch iron hoops. The capacity per vat is fifty-five tons of crushed dry ore. A new wooden 100 ton vat has lately been installed. The vats have a central discharge gate, 14x14 inches, discharging the tailings into launders underneath the tanks. The new vat is of larger size.

Below the leaching vats are situated the two gold solution storage tanks, ten feet in diameter and six feet deep, also of Oregon fir, with two and one-half inch staves.

Below the gold solution storage tanks are placed the two sump tanks, of the same size and kind as the above. These tanks were used formerly for precipitation tanks, employing zinc dust as a precipitant. This, however, has recently been discontinued.

The two solution storage tanks are situated above the leaching vats, and are ten feet in diameter and ten feet deep, with two and one-half inch staves. The main solution pipe lines in the mill are two and one-half inches in diameter, each leaching vat being connected, as is customary, with both the strong and weak solution gold storage tanks. There is no vacuum connected with the leaching vats, as the coarseness of the ore permits of rapid leaching and good drainage. Precipitation is carried on in wooden barrels by means of zinc shavings. There are five barrels for the strong solution and six barrels for the weak solution, the barrels being three feet deep and two feet in diameter at the middle. They are provided with a screen at the bottom, on which the zinc rests, and a rectangular compartment on one side, 8x4 inches, by which solution descends below the screen, rising up through the zinc as in the ordinary form of precipitation boxes. The advantages claimed for the barrels are: First, that the round shape tends to avoid channeling, which causes irregular precipitation, and, second, that the clean-up is much facilitated, because the barrels can be handled individually.

The Cyaniding of the Ores.

The leaching vats are charged by means of a belt conveyor, which takes the ore from the finished product bin and carries it to the five leaching vats. This belt conveyor was installed recently, the mill as originally designed having the vats filled by end dump cars running on tracks over the vats. The conveyor is in two parts, that is, there are practically two conveyors placed at right angles to each other. This is necessitated from the fact that the vats are placed in a row at right angles to the discharge end of the bin. The conveyor proper is a sixteen inch rubber belt, the first conveyor fifty feet long and the second, forty feet long. The speed of the belt is 600 feet per minute. By these means a vat is charged in from two to two and one-half hours, accomplishing a great saving in time over the old method of charging by cars. The tailings are discharged from the vats by sluicing with a water pressure of sixty to sixty-five pounds per square inch. The time needed to discharge a vat is two and one-half to three hours. The favorable situation of the mill gives unlimited grade to carry off the tailings.

While the vat is filling, and about one-half full, the strong solution is turned on, six pounds of Potassium Cyanide per ton of solution. The amount of strong solution run on is about fifteen tons, which is allowed to stand three to four hours generally, sometimes much longer, according to the nature of the ore. When the strong solution is started draining, the weak solution is run on and is allowed to remain undisturbed in contact with the ore for at least one hour; then leaching is started, fresh weak solution being continually added to replace that which drains off, until wash water is applied. The weak solution contains two and one-half to three pounds of cyanide per ton. The amount of weak solution thus brought into contact with this ore is from forty to fifty tons. When the gold value of the weak solution does not run above thirty to fifty cents per ton, the wash water is applied. The amount of wash water is about six to nine tons. The total time consumed is about ninety-six to one hundred hours, including the charging and discharging of the tanks. The total amount of the solution used, including the wash water, is seventy to seventy-five tons per vat.

The amount of lime used to neutralize the acidity of the ore and give the requisite protective alkalinity to the solution is six pounds per ton of ore. This is added unslacked on the cars, before the ore is dumped into the first bin.

The protective alkalinity is afforded by two and one-half pounds of lime per ton of solution.

The consumption of cyanide is 0.63 pounds per ton of ore, and the actual extraction ranges between eighty and eighty-five per cent. of the values contained in the ore.

The Precipitation by Zinc Thread and the Clean-Up.

The precipitation by zinc is practically perfect, but little trouble of any kind being experienced; the rate of flow of both weak and strong solution through the barrels is three tons per hour. The tailings solution issuing from the barrels runs from four to ten cents per ton.

The barrels are all provided with a strap on each side, so that they can easily be lifted. When ready for the clean-up the barrels are hoisted up and dumped into one of the gold solution storage tanks which has been emptied for the purpose. The men get right into the tank and wash the precipitates, taking out all the material that will not pass a ten mesh screen. This is put back into the barrels. The fine material in the tank is then pumped through a Johnson filter press, having 2x2 foot frames, which was formerly used in the zinc dust precipitation. The precipitates are thoroughly washed in the press and then partly dried by forcing air through the press. Then the precipitates are discharged and are ready for the acid treatment.

A small Worthington duplex pump operates the filter press by water pressure. An air pump can also be connected, so as to force air through the filter press. The precipitates, after the drying, are charged into the acid tank, which is 4x4 feet in cross section and five feet deep, lined with heavy sheet lead. It is provided on the inside with several coils of one-half inch lead pipe, carrying steam to heat the acid if necessary. The bottom of the tank has a slight slope toward the discharge opening, a one inch lead pipe in the bottom. The acid tank is provided with a close covered hood, which can be raised and lowered, to facilitate charging. The precipitates are treated with dilute sulphuric acid, of the usual strength, long enough to dissolve the zinc, then allowed to settle and the solution syphoned off. Then water is added and the mixture is pumped back through the filter press. Here it is again thoroughly washed with water, then dried with air. The dried precipitates are then heated to a dull red in sheet-iron pans, 2x4 feet in size, in a large iron muffle, then smelted in graphite crucibles with the following flux: Seventy parts crude borax, thirty-five parts soda and twenty parts of sand. The consumption of zinc is 0.9 pound per ounce of bullion produced.

Methods of Sampling Ore, Tailings and Solutions.

The pulp is sampled at the finished product bin by the following device: The belt of the conveyor has one hole on the center line of the belt, and every time this comes under the discharge spout of the bin some of the pulp shoots through, falling on the apex formed by two slant boards coming together under the center line of the belt. This divides the material into two samples, which are collected on the floor or in large boxes. No information is available as to the accuracy of the sample thus obtained.

SAMPLING THE TAILINGS.—The tailings are sampled by pipe sampler, four samples being taken irregularly distributed over the surface of the vat. After sluicing has commenced a face sample is taken at the center of the vat.

SAMPLING THE SOLUTIONS.—The solution pipes entering the gold solution storage tanks and leaving the zinc boxes are tapped by cork and a small rubber tube. No clamp is used, however, and the tube dips to the bottom of a five gallon stone jar, the overflow going back into the tanks. When the gold solution tank has been filled, the solution man dips to the bottom of the jar with a dipper and fills a small bottle, which goes to the assay office. The jar is then emptied and placed for another sample.

The Power Plant of the Mill.

Steam is furnished by two Erie return tubular boilers, seventy-five horse power. The engine is a Sioux Corliss, 12x36 inch cylinder, eighty-five revolutions per minute, steam at eighty-five pounds per square inch. A separate small upright engine furnishes the power for the conveyor and for the dynamo for electric lighting.

The Number of Men Employed in the Mill.

DAY SHIFT.
1 Crusher man.
1 Roll man.
1 Engineer.
2 Tank men.
1 Solution man.
1 Millwright.
1 Superintendent.
1 Assayer.

NIGHT SHIFT.
1 Roll man (takes rolls and crusher.)
1 Engineer.
1 Solution man.

Total twelve men. Wages for engineers $3.50 per day; other mill labor, $3.00 per day.

The Cost of Cyaniding.

The cost of cyaniding is $0.832 per ton, in detail as follows:

Labor	$0.411
Cyanide	.188
Zinc	.031
Repairs	.086
Fuel	.116
Total	$0.832

Other mills of the type described are those of the Spearfish and Deadwood Standard Companies, situated in the Ragged Top District. Wherever the ore yields its values to coarse crushing this type of plant is the most economical as regards cost of treatment.

A typical plant of the "wet crushing type" is that of the

DAKOTA MINING AND MILLING COMPANY IN DEADWOOD.

This cyanide plant is situated in the First Ward of Deadwood. It was constructed during the summer and fall of 1901 and went into commission in October. The mill is a wet fine crushing mill, having a capacity of 100 tons in twenty-four hours.

The Crushing of the Ore.

The ore goes from the bins, which are filled from railroad cars, to a Gates No. 4 D crusher, passing, however, first over the grizzley. The crushed ore is carried by elevator to the main storage bins, from which it is fed by suspended Challenge feeds to six five-stamp batteries. The weight of the stamps is 900 pounds, the crushing being done through a twenty mesh woven wire screen. The mortars have a double discharge, and instead of being the ordinary lip mortar are provided with a gutter below the screens, cast as one piece with the mortar. These gutters take the pulp discharge from the batteries and send it to the charge box of the spiral pump. A heavy canvas is hung over the screen and gutter to prevent splashing.

Four pounds of lime per ton are added at the battery to aid in the subsequent settling of the slimes, there being but little acidity present. The ore is crushed with a 0.1 per cent. or two pounds of cyanide per ton solution, known as "battery solution."

It has been found necessary to close up some of the rear discharges of the batteries in order to keep the amount of battery solution down to the normal amount.

The Separating of the Sands From the Slimes.

The pulp from the batteries is lifted by a Frenier sand pump (capacity 6,000 gallons per hour) to the launders feeding the separator boxes. The height of lift is about seventeen feet. The sands are separated from the slimes by means of classifier cones, two in series. The slimes in the overflow go to the slimes vats, while the sands are discharged at intervals into the vats directly below the cones.

The Tank Capacity of the Mill.

There are eight leaching vats, built in four tiers of two each, for the double treatment of the sands. The vats are of red cedar, twenty feet in diameter and five feet deep.

There are six slimes vats, built in two rows of three each, each three vats comprising a unit for the treatment of one lot of slimes. The vats are twenty

feet in diameter and twelve feet deep, two and one-half inch staves, and built of red cedar. There are two sump tanks, sixteen feet in diameter and six feet deep; two gold solution tanks, one twenty feet in diameter and eight feet deep, the other ten feet in diameter and five feet deep. Then there are two weak solution storage tanks, sixteen feet in diameter and twelve feet deep, and also a strong solution storage tank, twelve feet in diameter and seven feet deep. All these tanks are of red cedar. There is one tank twenty feet in diameter and eight feet deep used as a settling tank to settle out sediment from the siphoned slimes solution.

The Treatment of the Sands.

The sand vats are charged in about twenty-four hours, by means of the cones, and the pulp is then leached with a 0.2 per cent., or four pound cyanide solution, followed by the weak solution of two pounds per ton. The tailings are discharged by sluicing, the time consumed in discharging being about two hours. The total time consumed in treating sands, including charging and discharging vats, is five days. The extraction made on sands varies with the different ores treated, averaging about eighty-three to eighty-four per cent.

The Treatment of the Slimes.

The slimes, coming from the separator boxes, flow into one of the two lowest slime vats. When one of these vats is filled (the filling being done by carrying the slimes to the bottom of the vat by a wooden pipe), the siphon is started and the solution siphoned off, fresh slimes pulp meanwhile running into the vat continually. This can be done, as the slimes settle very rapidly with most of the ores treated at the mill. When so much slime has been charged into the vat that the siphoned solution begins to be seriously discolored, twenty-five pounds of slacked lime are fed into the launder taking the slimes overflow from the separator boxes, and when this has been carried into the slimes vat, the pulp is changed into the lowest slime vat of the second unit, where the above operation is repeated. The slimes in vat No. 1 are then allowed to settle for one hour, when the siphon is again started, and now follows the settling of the slimes down until the solution has been drained within a few inches of the settled slimes. The total time of settling is about ten hours. The settled slimes in this vat are now sluiced out with "barren solution" to the centrifugal pump, which lifts the material to the next slime vat of the unit. When the vat has been filled the pulp is drawn off at the bottom by the centrifugal pump and returned to the vat over the top. This circulation is carried on for an hour. Then the slimes are allowed to settle, the clear solution at the top being siphoned off during the settling. When the slimes have settled, fresh barren solution is added and they are circulated by the centrifugal pump for another hour, twenty-five pounds more of lime being added in the course of the pumping. The slimes are then again allowed to settle, while the solution is siphoned off as before. When settling is complete the slimes are sluiced out with wash water to the centrifugal pump, which now raises the pulp to the last slime vat of the unit. Here the slimes are allowed to settle again and the solution siphoned off. After the solution has been drawn off wash water is added by a nozzle and pipe device situated in the vat. This serves to thoroughly wash and agitate the slimes for the last time. The slimes are now allowed to settle, while the solution is drawn off by the siphon. Then the slimes are sluiced out to waste, having passed through three vats. The extraction made on slimes is about eighty-eight per cent., but at present only about eighty to eighty-two per cent. is saved, the balance passing out in the tailings in the retained moisture. More thorough washing, with facilities for handling this increased bulk of wash water, will undoubtedly raise the percentage saved. The total time consumed is about six days.

The Precipitating Department.

Precipitation is carried on in four zinc boxes, two for siphoned solution and two for sands solution. There are ten compartment boxes, each compartment having a cross section of 20x15 inches and a depth of about eighteen inches. Some extra precipitating barrels are being put in to accommodate an increased amount of solution. The precipitates can be sluiced from the boxes to the acid treatment tank, where they are washed and treated with dilute sulphuric acid in the usual way, transferred to a filter tank, filtered, washed and dried, and then smelted. The slags are treated at the Golden Reward Smelter.

The precipitation in the boxes is good for both sands and slimes solutions, the slimes solutions usually carrying from .05 to .08 per cent. free cyanide. Considerable sediment sometimes accumulates in the boxes. but does not seem to give any trouble in precipitation.

The Nature of the Ore Treated.

The ores treated at the mill come mainly from the vicinity of Portland, and are siliceous Potsdam ores, with the values very finely distributed. The mill also treats custom ores to a considerable extent. Recently some siliceous ores have been treated which were but slightly oxidized, seemingly with considerable success.

The Dakota Mill is very similar, at least in plan of treatment, to the Mogul Mill of the Horseshoe Company at Terry, the Penobscot Mill at Garden City, the Columbus Mill at Central the Hidden Fortune Mill and other smaller mills of the district. The cost of treatment by the method described is from $1.25 to $1.50 per ton, depending on the scale of operations.

A typical plant of the third type (fine dry crushing) is that of the

IMPERIAL MINING AND MILLING COMPANY OF DEADWOOD.

This plant is located a short distance from the Dakota plant, already described. It is a dry crushing plant, with roasting facilities for the heavier 'blue" ores, and has a capacity of 150 tons per day, with provisions to increase to 250 tons in time. The power plant is sufficient for the latter capacity. The plan of the mill follows the Colorado pattern at Florence, treating telluride ores from the Cripple Creek district. It is built entirely on level ground, the crushing, roasting and leaching departments being in separate, but closely adjoining buildings, the transference of pulp being accomplished by elevators and belt conveyors. The whole plant is built very substantially on solid concrete foundations. The buildings are all iron clad.

The mill represents a different type from that usually to be found in the Black Hills, and as it is to treat ores of the same nature that the Dakota and Portland mills are treating, in fact from properties in close vicinity to the mines of the latter mills, a comparison could be made between the two methods of treatment, which would be of considerable interest, as regards the relative merits of fine dry and fine wet crushing with the slimes treatment of the latter method. As the mill is designed to treat custom ores, two Vezin samplers, placed in series, are being installed, with the necessary sample crushing plant to further reduce the sample cut out by the samplers.

The Crushing Machinery.

A shaking grizzley, four feet long by two feet wide, takes out the fines, while the coarse ore goes to a 10x20 Blake crusher. The ore is further crushed by one set of Davies rolls, sixteen inch face and thirty-six inches in diameter, with Latrobe steel shells. The rolls make sixty revolutions per minute. The crushed ore then goes through the Vezin samplers, after which the bulk of the ore is dried in a Davies multi-tubular dryer (a revolving cylindrical drying furnace), and then goes to two sets of finishing rolls after being screened over an eight mesh inclined screen. Each finishing roll, with its following four revolving screens and elevators, is constructed as a unit, so that one set of rolls can be run independently of the other. The advantage of this is evident in case of a breakdown in the crushing machinery. The finishing rolls are of the Davies type, thirty-six inches in diameter and sixteen inch face, having a speed of eighty revolutions per minute. It is seen that the roughing rolls and finishing rolls are of the same size, permitting of change of the roll shells, from the finishing to the roughing rolls, when the first become too uneven for uniform crushing.

All the rolls and screens are steel housed, and these housings are connected by two Sturtevant exhaust fans, with two Prinz and Rau dust collectors, consisting of a revolving spindle, on which are stretched collecting bags. The dust is drawn in at the center of the spindle and collected in the bags, from which in turn it is discharged, by automatically shaking the bags, the collected dust being carried by a conveyor to the pulp bins and mixed with the pulp. In this way no dust is accumulated for separate treatment, but what is formed is treated with the pulp.

The pulp storage bin is divided into two divisions of 250 tons each, one for sulphide ores and one for the oxidized ores, these being kept separate throughout their course through the mill, the sulphide ore going from the bins to the roasting furnace.

The Roasting Furnace.

This is a Holthoff-Wethey two hearth furnace, the upper hearth being the roasting hearth, while the lower one acts as a cooling hearth. The inside dimensions of the hearth are 121 feet length and twelve feet width. The outside dimensions are, length 131 feet, width fourteen feet. By the time the ore gets to the lower end of the cooling hearth it is cool enough to be taken by the elevator and raised to the bins. The ores to be roasted are light in sulphur.

The Leaching Department of the Mill.

There are four steel leaching vats, thirty-five feet in diameter and six feet deep, the body being of three-sixteenth inch steel and the bottom of one-fourth inch steel. The vats hold 200 tons and have five bottom discharge gates, operated from above. There are also four smaller leaching vats, placed so as to utilize the waste room left by the large vats, due to their circular shape. These are ten feet in diameter and six feet deep, holding sixteen tons. There are two gold solution storage tanks, sixteen feet in diameter and four feet deep, and two sump tanks twenty feet in diameter and four feet deep and two solution storage tanks, twenty feet in diameter and four feet deep, all of steel, same size of material as the leaching vats. Precipitation is carried on by zinc thread in steel precipitation boxes. The sulphuric acid method is used to refine the precipitates. The leaching vats are connected with vacuum chambers, to facilitate the final washing of the pulp.

The Power Plant.

There are two Heine water tube boilers, of 125 horse power each, to carry 150 to 200 pounds of steam pressure. The engine is a 250 horse power tandem compound Hamilton Corliss, non-condensing. There is an electric light plant of 250 lights. The power plant is large enough for a 250 ton mill, and it is intended to eventually install two more sets of finishing rolls, with their screens and elevators, and four more thirty-five foot leaching vats. The actual cyanidation of the crushed ore is very similar to that described for coarse dry crushing.

(We add in connection with Mr. Fulton's article a description of the Lundberg, Dorr & Wilson plant by Mr. A. D. Wilson, as being a new type of mill for the Black Hills and which had not been completed when Mr. Fulton's article was written.—Committee.)

LUNDBERG, DORR & WILSON CYANIDE MILL.

This mill, of seventy-five tons daily capacity, situated upon the property formerly owned by the Buxton Mining Co., is held under the private ownership of John Lundberg of Terry, S. D., John V. N. Dorr of Terry, S. D., and A. D. Wilson of Deadwood, S. D. This is the first mill in the Black Hills to use electric power, to use a modern type of the Chilian mill, to use a belt elevator for the elevation of pulp, to use the Moore Process for the finished treatment of slimes, and to use the Dorr Mechanical Classifier. The ore first passes through a No. 4 Gates crusher, and is elevated to a storage bin. From the storage bin it is fed by a shaking trough to a set of coarse rolls which reduce it to about three-fourths inch size. As it passes through these rolls the solution of cyanide of potassium is added and carries all the material to the Chilian mill. This mill is a six foot Monadnock Mill, manufactured by S. V. Trent & Co., of Salt Lake City, whose three ponderous rolls, weighing 6,600 pounds, revolving at a speed of thirty-one revolutions per minute, strike ninety-three blows per minute upon any part of the die. From the mill the ore and solution is elevated by means of a belt elevator. to the Dorr mechanical classifier, where the sands and slimes are separated. The sands are carried from the classifier to the leaching tanks, eighteen feet in diameter and ten feet deep, where they undergo the ordinary process of leaching. From the classifier the slimes are carried to the cone settler. This cone is twenty-two feet in diameter and fourteen feet deep, with a carefully leveled overflow. The clear solution is reused and the thickened slimes are drawn from the bottom and delivered to the slime storage tank where they are given an aeration. From the slimes storage sank the slimes are drawn off to the first tank of the Moore Slimes Process. This tank has hopper bottoms and air pipes. From the bottoms of the hoppers the slimes are drawn continuously by means of a power diaphragm pump and poured back into the top. This, with the agitation provided by the air pipes, gives the slimes a thorough aeration.

The Moore filter plates of which there are thirty-four 4x6 feet, attached to one framework, affording a filtering surface of 1,632 square feet, are then lowered into this first tank, and the vacuum pump started. The solution comes through these filters very clear and is piped to the zinc boxes. The slimes form a thick coat upon the outside of the filter frames. When a thickness of about three-fourths inch is attained the filters are raised out of the tank and by means of a traveling crane are moved over and immersed in a tank carrying barren solution. The suction is still continued and the gold bearing solution replaced by the barren solution. The filters are again raised and immersed in clear water, which replaces the barren solution. The filters are again raised and moved over the discharge hoppers where a large volume of compressed air is blown into them. This releases the coating of slimes and carrying about thirty per cent. moisture it drops down into the discharging hoppers from which it is drawn into a hopper bottom car and thrown out on the dump. The length of time occupied by the above cycle is three hours, and the amount of slimes to a charge is four and one-half tons dry. The sands are shoveled from the tanks through a central discharge hole into a four ton car, with side discharges, and thrown out on the dump. The mill is under the superintendency of John V. N. Dorr.

WET CRUSHING IN SOLUTION.

By John M. Henton.

[Paper read before the Black Hills Mining Men's Association, April 22, 1903.]

My subject is "Some of the History of Wet Crushing in Cyanide Solution," and so you are guarded against expecting all the history and I can excuse myself for all omissions. What I may say purports to be history, and history is a record of events which should teach something by showing the steps of progress and development, the details of which are often uninteresting and wearysome and I ask your kind indulgence.

By way of introduction I will quote from "The Cyanide Process, Its Practical Application and Economic Results," by Dr. A. Scheidel, San Francisco, October 1, 1894. On page 28, in writing of the practice in general Dr. Scheidel says, "Many attempts have been made to discharge ore pulp direct from the mortars into the percolating vats but their successful treatment by cyanide, when so discharged, has been prevented by mechanical causes and in consequence of the presence of slimes the results are unsatisfactory. The advantages of wet crushing over dry crushing are so obvious however, that experiments will be continued and ultimately the draw-backs which now adhere to the method will be overcome. Cyanide of potassium solution has been used, in some instances, and in an experimental way, in lieu of water in the mortars, when wet crushing has been resorted to, but does not appear to be practiced anywhere at present."

On page 48, speaking of the practice in South Africa, he says: "In an experimental way, a solution of cyanide has been used instead of battery water; at present, however, water is invariably used in the mortar boxes, sufficient success not having attended the other methods. The use of cyanide solution in the mortars would be of advantage only when the pulp is directly delivered into the percolation vats; the formation of slimes is fatal to this method." On page 76, an extract from the New Zealand Government Report is given: "The Tryfluke Co. tried to run the tailings directly into the tanks but they, like others, found that the amount of slimes in the ore prevented the cyanide solution from filtering." It will be noticed that one universal difficulty was encountered, namely, slimes. The word slimes is the key to a correct understanding of the subject of wet crushing in cyanide solution.

I will quote now from page 67 of Bulletin No. 5 of the South Dakota School of Mines. I make this quotation in order to correct and not to find fault. "The method of working the Deadbroke mill, treating the pulp from the plates, led to the direct crushing of the ores in the battery with cyanide solution in the Gayville and Central City mills." How such a result could have followed from the method of working the Deadbroke mill it is hard to imagine, for no attempt was ever made at that mill to treat the slimes which were allowed to run wholly to waste, and thus the slimes problem remained in the old unsolved condition as related to saving any values which they might contain; nor was direct filling attempted until after the building of the Dakota mill at Central City by Capt. Bullock; nor was cyanide at any time ever used in the batteries. And in any other way than those any other stamp-mill could be said to have led to this method of cyanide practice. The slimes problem being the only serious difficulty to meet in wet crushing in cyanide solution, it might have led to a better understanding to have entitled this paper "Some of the History of Slimes Treatment."

February 19, 1898, Chas. Butters, a chemist of South Africa, delivered an address in Johannesburg on the treatment of slimes which gives the best practice up to that date.

The reading of the address would take an entire evening and only a few sentences can be given here. After saying that he had "been specially employed on the slimes problem for the past three years," he continues, "These researches have resulted in the working out of a practical method which consists in the coagulation of the slimes in battery water by means of lime, their concentration by spitzkasten and final settlement in continuous overflow vats. These settled slimes are now ready for treatment. Three methods for dissolving the gold are at present in use on these fields. One consists for pumping the fresh slimes from one tank to another through a centrifugal pump; the second, the fresh slimes are agitated in a dissolving vat, by means of a stirring gear, for six or eight hours; the third consists of a joint use of both of these former methods. At the Bonanza, agitation, aeration, with an air compressor through a pipe placed near the bottom of the vat, and circulation through a centrifugal pump, are all in use at the same time. With decantation the washing of the material takes place by dilution, one wash does not displace another wash but simply mixes with it. Decantation as ordinarily practiced means that a ton of solution is thrown away with every ton of slimes. To reduce this large amount of solution thrown away we have erected at the central works of the Rand Central Ore Reduction Company, the two largest vats in the world, each being fifty feet in diameter and sixteen feet deep. It has been found that by filling these gradually and allowing them to settle for about a week, the pressure exerted by the great depth of settled slimes reduce the average moisture to about forty per cent." Mr. Butters closes his address by saying, "It appears possible that the aim of the metallurgist will soon be to increase the percentage of slimes produced by the mill."

Americans are not generally much behind the balance of the world Mr. M. W. Alderson writes of what was being done in Montana in the M. & S. Press, "In the winter of 1895–6 laboratory tests were made in Bozeman, where extractions as high as ninety-six per cent. were made in a quarter of an hour with a cyanide solution. A company of Bozeman men arranged with the owners to handle the tailings (at Iron Rod) on a percentage of their assay value. F. W. Traphegan, Ph. D., went to work to elaborate a scheme whereby the material could be handled, and decided on a combined system of agitation and percolation, the slimes to be coagulated by the use of lime during agitation, when percolation could be carried on without difficulty. The agitator was built, a charge put in, material was flocculated and everything worked nicely until attempt was made to draw the fluid. Every affort to secure percolation of the charge resulted in absolute failure." Mr. Alderson continues, ''I was impressed with the fact that the moment agitation ceased the sand and slimes sought the bottom of the vat and the fluid rode the top. I remarked that every successful invention was the result of working with nature, and that successful handling of slimes would come about from drawing the fluid from the top, where it naturally came. The percolation was discontinued and decantation was attempted, and the mechanical difficulty in the way of treating the slimes soon overcome." You will notice that Montana men, in the work of solving the slimes problem, were contemporaneous with South African chemists, and did not copy from them.

During 1897 and 1898 I was in the employ of Mr. Alderson in his cyanide works in Montana, and had the benefit of all his varied experience at a number of places. In 1899 Capt. Bullock employed me to make experiments on the Gunnison ores to determine their amenability to cyanide and the work was successful enough to warrant the building of a mill. Mr. Bullock, as manager for the Dakota Mining and Milling Company, said to me that there was an old stamp mill in Central City he could get. I told him that a stamp mill was my choice, and so it was determined to fit it up for use, and I drew the plans. Mr. Bullock gave to me his confidence, and the difficulties were freely discussed as they occurred and were suggested to him, and always when he said slimes, which he frequently did, I quoted Butters and said, ''The more slimes the better." My experience in Montana had led me to think with confidence of being able to treat them successfully. But I never mentioned that I intended to use cyanide solution in the mortars, for I was afraid he would veto it, and I believed it to be the key to the successful treatment of the ore.

Mr. Bullock would generally end our discussion of ways and means by saying, "Well, I believe you know what you are doing and I will leave it all to you." In determining to use cyanide in the mortars I had but one recorded success to give me confidence. Mr. Frank Merricks in a paper read at a meeting of the Institute of Mining and Metallurgy, London, this same year, 1899, said, ''About eighteen months ago, at the Crown Mine, New Zealand, wet crushing experiments were made with cyanide solution passing through the mortars. The method of handling the slimes and the results obtained, have not been made public, but are presumed to have been satisfactory as the plant has been changed to wet crushing." This is all that was said regarding the method. Mr. Alderson had not used this method nor suggested its use. The mill at Central City was built on level ground and the plan was to catch the pulp in a hopper and elevate it into a vat where it was stirred by a Alderson propellor blade agitator. The mill was started up early in October, 1899. I did not put cyanide into the mortars until the filling of the second vat. Mr. Bullock came in in a few hours afterward and one of the mill men told him. He came to me and said, ''You are putting cyanide into the batteries." I replied, ''Yes." Nothing more was said by either of us until the results were known. The extraction and saving were just as good from the very start as during any subsequent work. But there were mechanical difficulties to overcome in order to increase the capacity and secure economy.

One of the first of these was the method of lifting the pulp up into the agitator vat from the mortars. It had been planned to use a centrifugal pump, but some one suggested a steam jet. It so happened that just at this time the papers were full of the great efficiency of hot alkaline solution in the treatment of gold ores. It does seem plausible when we remember what scientists say on the subject. LeCoute says, ''water at 752 degrees F. reduces to a pasty condition nearly all ordinary rocks—and in the presence of alkali, even in small amounts, the same result is produced at as low as 300 degrees.'' Water boils at 212 degrees F., so you see how a great metallurgical result might be expected without using excessive heat, so much so that hardly a year passes in which we do not hear of some such attempt. I had no faith in a commercial success through such agencies for the reason that the alkaline solutions are to no considerable extent selective in their effects, and no metallurgical reagents which are not selective can be successfully applied. This proved true in this case, and before many days the solutions became so foul, so loaded with base salts, that a change was necessary; the expense of using steam in this way also made it impracticable, and a centrifugal pump was substituted. The first method of treatment tried was agitation of the entire battery product, which was then discharged into a second vat and the solution drawn both by percolation and decanting.

This proved all right in saving of the values, but the sands would at times be so heavy and packed so solid they could not be agitated, then the ore would change and the pulp be so light and slimy that nearly all the solution would have to be drawn off by decanting. To provide better for this varying condition another and larger vat sixteen feet by eight feet, was built and the slimes washed out into it, passing through the centrifugal pump, by an agitation overflow, and then the sands discharged into the percolation vat. The mill was an old antiquated pattern of the seventies, and short in power and when agitating, the batteries had to be hung up. It was seen that if the separation of sands from slimes could be made outside of the agitator and only the slimes treated in it, that the running time of the batteries could be nearly doubled. To accomplish this a V box was introduced and accomplished all that was expected from it. The sands thus separated were treated by the conventional methods. The slimes were treated in the Alderson agitator for three or four hours, then discharged through the centrifugal pump, with barren solution wash, into the big slime vat where settling and decanting took place. It all looks very simple as finally worked out, but before it was accomplished the management put in many a twenty-four hour shift and Capt. Bullock walked up to Central many times while the morning star was still shining.

It is not claimed that any of the work done represents the best practice, but that the results obtained were secured in the way stated.

I contributed to the Mining and Scientific Press of March 10 and September 8, 1900, articles giving full details of this work. On January 14,

1903, I wrote to the publisher saying that so far as I knew the above articles by me gave the detail of the first successful practice of wet crushing in cyanide solution published in the United States, and asking that if he knew of any prior successful work being done to mention it. In reply he wrote:

SAN FRANCISCO, CAL., January 20, 1903.

MR. JOHN M. HENTON,
    Deadwood, S. D.

Dear Sir:—Replying to your inquiry of the 14th inst., so far as we are able to investigate, the stand you take is in accordance with the facts. I have not yet been able to note any account of the real experiments in California referred to, but the subject will be given further attention. With personal regards,   Very truly yours,

J. F. HALLORAN.

Mill men engaged in dry crushing, where fine crushing is necessary, have the slimes problem to deal with and Mr. W. A. Watson has written to me a very interesting letter on the subject of treating this dust, which I take pleasure in giving a synopsis of:

DENVER, COLO., April 17, 1903.

MR. JOHN M HENTON,
    Deadwood, S. D.

Dear Sir:—Your letter under date of April 16th reached me this evening, and I am pleased to acknowledge receipt, and I wish to applaud the action of the Mining Men's Club of the Black Hills in getting together for discussion of ideas, and the presentation of views and practices. * * *

The old Golden Reward mill, as all Black Hillers know, was a dry crushing plant, and we had the usual, or possibly more than the usual amount of dust. * * * This was overcome to a large extent by installing an ordinary "Buffalo" exhaust fan, the same as is used in handling shavings and saw dust in a planing mill. This dust at first was run into a room where a jet of exhaust steam was turned. The collecting of the dust was remarkably well done in this way. Our idea was to make brick of this mass, and to ship it to the smelter. We did make a large quantity of brick; but when we learned of the treatment charge, the cost to brick and to load into cars we arrived at the fact that we "were up against it hard and fast." * * *

After the days of mud we went back to the dry dust, which was collected in the upper part of a large bin by means of triangular or trough-shaped screens made of large sheets of burlap. By arranging them at very acute angles we were able to get into our "bag house" a large number of square yards of burlap, and at the same time permitting these bags to free themselves of the collected dust, and to discharge into the hoppered bin below. The exhaust from this bag house was passed through a long horizontal dust collector three feet deep by four feet wide with hopper bottom and discharge slides which emptied into a car. This collector extended through twice the length of the mill, 140 feet, if I remember correctly. The whole collector was dust tight and within it was hung strips of burlap three feet wide extending the full length of the collector, fastened only at the top and the inlet end, resembling somewhat a book opened to separate the leaves and then suspended from the back. These leaves were arranged one inch apart, the inlet end being securely fastened, but the opening between each leaf free for the passage of air to enter. The air caused a waving motion of the burlap which arrested the remaining dust almost completely. Between the hoppers at short distances, a partition was extended up to the lower edge of the burlap. This was to compel the air to pass between the leaves and not under.

Our slime machines were two in number, and consisted of two steel tanks, thirteen feet in diameter by four feet deep. All of our machines and the general arrangements were built to comply with the available space in the mill building. These tanks would have been better had they been greater in diameter and three times the depth. Each tank had a hollow shafting which stood vertically in the center and extended about two feet above the top, and was fitted to a bevel gear which was driven by a belt. Each machine had its clutch for starting and stopping. Just under the driving gear were two cross arms from which were suspended by chains, sheets of boiler plate one-half inch by six inches by eighteen inches, which, when hanging vertically, reached within six inches of the bottom, and arranged upon the cross arms in such a manner that the whole contents would be thoroughly mixed. At the bottom of the hollow shaft were fitted four one-inch perforated pipes, each extending outward to the circumference of the tank. The perforations were one-sixteenth of an inch in diameter, and arranged about one inch apart. Each tank was arranged with an overflow box, which carried the clear supernatant liquid into the solution tanks.

A charge of three to four tons of dry slimes was dumped into the tank and enough cyanide solution of a strength 0.3 to 0.60 per cent. added to make the mass the consistency of thin mortar, the mixing paddles were lowered into the tank, and the whole mass thoroughly mixed and agitated from three to six hours. The mixing device was then raised and allowed to hang at rest. The hollow shaft continued to revolve very slowly and the charge allowed to settle a foot or more and then the weak cyanide solution was allowed to pass into the mass down through the hollow shaft and out the perforations in the lower pipes, under low gravity pressure. The richer solution was gradually displaced by the weaker, and overflowed and run directly to the gold solution tank, and from thence to precipitating boxes. The weak solution was followed by water. * * *

W. A. WATSON.

A number of bright, practical men are engaged in trying to improve the methods of slime treatment as practiced today, and we may expect that some one or possibly more, will ultimately make a marked success. The fact remains, however, that so far as relates to the mill practice as commercially carried on, but little advance has been made over that described by Chas. Butters. The economic difficulties in the treatment of low grade slimes are great and it will be a marvel if the Homestake company, for instance, succeeds in treating at a profit a $1.00 slime. It is the economy of the stamp mill as an ore pulverizer that gives to wet crushing in cyanide solutions the prominence it has received, and any new or improved method of slimes treatment must bring with it this same element of advantage, economy of operation and maintaining in efficient working order.

CYANIDING PRACTICE AT THE MAITLAND PROPERTIES.

By John Gross, Mill Superintendent and Chemist.

[Paper read before Black Hills Mining Men's Association, April 20th, 1904.]

INTRODUCTORY.

The group of claims, comprising over 1,100 acres, located at Maitland* in the Ida Gray Mining District, Lawrence County, South Dakota, owned by Alexander Maitland, are being developed and operated by the owner.

Prior to the acquisition of this property by the present owner, comparatively little work had been done and but little was really known of this district. Within the last two years, however, development has opened up some very promising ore chutes and the property is today in excellent shape.

The ores so far encountered are the so-called Potsdam siliceous ores of the flat formation lying on top of the Cambrian quartzite; both oxidized and blue, or unoxidized, ores are met with; a small amount of the blue ores are unavoidably sent to the mill, special care is taken, however, to send only the oxidized ores for the cyanide treatment; the sulphur in the mill ore will run from one to two per cent.

The ores are close grained and hard and have given quite a little trouble in their treatment. Pyrrhotite and iron pyrites exist in about equal proportions. Arsenic and traces of copper, antimony and tellurium are found and bismuth has been detected in the bullion. The silver contents predominate slightly over the gold in the low grade while the reverse is generally true in the high grade ores.

The following analyses give a fair idea of the general character of these ores:

Oxidized ores from two different mill samples, covering a period of several months.

	No. 1	No. 2
Au.	0.56 Ozs.	0.69 Ozs.
Ag.	1.03 Ozs.	1.50 Ozs.
SiO_2	70.95%	73.20%
Fe.	10.30%	10.40%
S.	1.66%	0.63%
As.	0.30%	0.00%
Sb.	Trace	0.00%
Te.	0.002 %	Trace
Cu.	0.02%	0.004%
Zn.	0.00%	0.00%
Mn.	Trace	0.75%
Al_2O_3.	4.30%	2.14%
CaO.	3.40%	3.20%
MgO.	1.02%	Not Determined.

*Formerly known as Garden City.

Blue ores from two general samples:

	No. 1	No. 2
Au	0.63 Ozs.	0.85 Ozs
Ag	2.00 Ozs.	6.08 Ozs.
SiO_2	65.38%	80.00%
Fe	13.40%	7.50%
S	11.40%	4.40%
As	0.90%	2.00%
Sb	Trace	0.00%
Te	0.003	Trace
Cu	0.02%	0.004
Zn	0.00%	0.00%
Mn	Trace	0.54%
Al_2O_3	5.43%	1.79%
CaO	2.10%	1.70%
MgO	0.20%	Not determined.

High grade ores, No. 1 being a blue and No. 2 a brown ore:

	No. 1	No. 2
Au	3.35 Ozs.	2.00 Ozs.
Ag	1.75 Ozs.	0.62 Ozs.
SiO_2	80.90%	84.80%
Fe	9.94%	7.50%
S	4.53%	0.75%
As	0.29%	0.00%
Sb	Trace	0.00%
Te	0.007%	Trace
Cu	0.013%	0.008%
Zn	Trace	0.00%
Mn	Trace	0.96%
Al_2O_3	1.70%	1.02%
CaO	0.50%	0.90%
Mgo	Trace	Not determined

Early in 1902 the building of a forth stamp wet crushing cyanide mill was begun and finally placed in commission by January 1st, 1903. The treatment, as outlined, was to crush in a cyanide solution, separating the sands from the slimes and the treatment of the slimes by agitation and decantation.

The ore is trammed from the shaft, 300 feet east of the mill through a covered tramway and delivered to the crusher bin.

The boilers are all consolidated near the hoist, while nearby a well equipped machine and blacksmith shop are located.

A 50,000 gallon water supply tank on the hill back of the mill supplies the necessary water for the mill and boilers and is obtained entirely from the mine workings.

MILL PRACTICE.

Accompanying this paper is a metallurgical plan of mill showing the complete treatment of the ore.

All figures given in this paper are based on a daily tonnage of 100 to 110 tons as now maintained.*

---

*Since this paper was written the capacity of the mill has been increased to very close to 120 tons per day.

CRUSHING THE ORE.

The ore as it comes from the mine, carrying an average of eight per cent. moisture is delivered to crusher bin of 150 tons capacity.

The crusher is a 24x13 inch Blake running at 260 revolutions per minute and is set to crush to a size that will pass through a one and one-half to two inch ring. The average time of running the crusher is seven hours per day equal to a capacity of fifteen tons per hour.

We have no figures at hand on costs of iron used on the crusher as none of the plates have been worn out.

The crushed ore passes to the elevator where it is elevated forty-four feet for delivery to the battery bins.

In the original mill installation a continuous type bucket elevator traveling 100 feet per minute was provided but was soon discarded after handling 9,300 tons of ore; the links had worn out causing the elevator to fall into the pit on several occasions.

A fourteen inch eight-ply rubber belt is now used, traveling 350 feet per minute; to give longer life to the belt it is now reinforced between the buckets with old pieces of belt which take the roughest wear, but one belt has so far been worn out after having handled 16,965 tons at a cost of 0.91 cents per ton ore. Twelve inch buckets are spaced eighteen inches apart on the belt. No. 10 steel buckets were first used and handled 16,965 tons at a cost of 45 cents per ton ore.

Malleable iron buckets proved to be too light for the work and handled 8,080 tons at a cost of 0.76 cents per ton. We are now using No. 6 steel buckets but have no figures on these yet.

The ore as it falls from the elevator head to the battery bins is cut by an automatic sampler which cuts out one-fiftieth of the ore and delivers it to the sample room where it is cut down daily.

STAMPS.

The battery bins, of which there are two of 150 tons capacity each, are located one behind each set of twenty stamps. Hung challenge feeders are used and are giving very good satisfaction.

The mortar is narrow, single issue type, and a six inch discharge is maintained by means of chuck blocks. Thirty mesh, No. 28 wire screens were originally used, but have been replaced by 26x13 mesh, No. 26 wire; this screen giving a longer opening does not choke so readily; the screens used are rolled, which also helps to keep the holes open. The life of one of these screens is about fifty days.

The stamp weights are as follows:

	Pounds
Shoe	150
Boss head	250
Stem	375
Tappet	135
Total weight of stamp, new	910

Ten cams with Canda fasteners are on one shaft.   Ninety-seven drops per minute of seven to eight inches are given, the order being 1–3–5–2–4.

Chrome shoes and dies are now being used but other makes have been and are to be experimented upon.

A Chrome shoe wears from ninety to ninety-five days, crushing 250 tons of ore at a cost of 4.90 cents per ton.   A cast iron shoe wears from thirty-five to forty days crushing 105 tons of ore at a cost of 4.95 cents per ton.

We have no records as yet on Chrome dies.   Cast iron dies wear forty days, crushing 105 tons ore at a cost of 3.28 cents per ton.   Wilson forged steel dies wear 105 days, crushing 280 tons ore at a cost of 3.06 cents per ton.

The above figures on shoes and dies cover a period of eight months running.

The battery solution is kept at a strength of 1.2 to 1.3 pounds of KCy and a protective alkalinity corresponding to from 0.8 to 1.0 pounds of NaOH per ton.*

The lime is fed with the ore into the battery.   An analysis of the lime used is given below:

$$CaO. \dots\dots\dots\dots\dots\dots\dots\dots\dots\dots\dots\dots\dots\dots\dots\dots\dots 92.0\%$$
$$MgO. \dots\dots\dots\dots\dots\dots\dots\dots\dots\dots\dots\dots\dots\dots\dots\dots 0.4\%$$
$$Al_2O_3+Fe_2O_3. \dots\dots\dots\dots\dots\dots\dots\dots\dots\dots\dots\dots 1.2\%$$
$$Insoluble . . \dots\dots\dots\dots\dots\dots\dots\dots\dots\dots\dots\dots\dots\dots 2.4\%$$
$$H_2O+CO_2. \dots\dots\dots\dots\dots\dots\dots\dots\dots\dots\dots\dots\dots\dots 1.1\%$$

The stamp duty is not high, for the last six months of 1903, the average was 2.66 tons; this has now been increased very close to three tons.   The amount of solution going to the battery is between four and five tons to one ton of ore; this, together with the hardness and compactness of the ore accounts largely for the small stamp duty.

The stamps deliver a product carrying about sixty per cent sands, and forty per cent slimes.   The material that we call slimes is that portion of the ore that will make water muddy; sands, no matter how fine, will not cause water to appear muddy.

The loss of time on battery running for the last seven months, when the mill started working full shift has been 5.8 per cent.

SEPARATION.

The separation of the slimes from the sands is one of the most vital, if not the most vital one, in the wet crushing process and several systems were tried but found to be inadequate.

In the separating system that we are at present using we have found it to be advisable to make a clean sand rather than a clean slime; a sand charge with five per cent or higher in slimes giving us a low leaching rate so that we are making this clean sand at the expense of throwing some sands into the slimes; these sands are, however, very fine and cause no

---

*But two solutions are used in the mill, the battery solutions as noted above, assaying about 50 cents per ton, and the barren solution with a strength of 1.5 to 1.6 pounds KCy and a protective equivalent of 1.0 to 1.2 pounds of NaOH per ton of solution.

trouble in the slime department and assay after treatment the same as the slimes proper.

The pulp that the battery delivers, flows to two Frenier sand pumps 54x10 inches, making nineteen revolutions per minute, and is raised twenty and one-half feet to a box provided with a screen to catch any foreign substance, thus preventing the choking of the cones; this box delivers the pulp to two upper cones and is intended primarily to take care of the intermittent discharge of the sand pumps, giving a more steady feed to the cones.

These two upper cones are simple in construction, forty-two inches in diameter at the top and having vertical sides for twelve inches down, at which point the cone starts at a sixty-degree slope, ending in a six-inch diameter sorting column with a two-inch discharge at bottom provided with a cock.

The sands, containing twenty-five to thirty per cent slimes, discharge at the bottom of the two upper cones, combine and flow to a single cone, the same size as the upper ones, but provided with an upward current of solution. This upward current is taken from the battery solution stock tank.

The sands discharging at the bottom contain from one to two per cent of slimes and go direct to sand vats.

The slimes overflow from the three cones contain from fifteen to twenty-five per cent sands of which only a small portion will stay on a 150 mesh screen.

The products going to sand and slime vats from the cones amount each to very close to fifty per cent of the original ore; the average since the cone system was installed, eight months ago, has been 48.2 per cent to sand and 51.8 per cent to slime vats.

SAND TREATMENT.

The clean sands from the lower cone issuing with 2.8 parts of solution to one part sand flow through launder with a grade of seven in 100 to the distributor over the sand vats. The amount of solution with the sands coming from the cone is not sufficient to carry same through launder and keep the distributor open and running; to overcome this, sufficient solution is added in the launder to bring it up to at least five of solution to one of sands.

The distributor is of the Butters type, ball and roller bearings, with six arms.

The sand vats, of which there are six, are thirty feet in diameter by six feet deep, having a lattice filter frame; eight-ounce duck cloths are used on top of cocoa matting; the eight-ounce duck has been found to be more satisfactory than the heavier grades.

The sand vats hold 140 tons and are filled in about sixty hours.

The system of filling the sand vats through a vat full of solution has been discarded in favor of dry filling, *i. e.*, the vat contains no solution when starting to load and all incoming solution with the sands is allowed to drain off as rapidly as it enters, keeping practically a dry surface on top of the sands ; this gives a better leaching product as the slimes with the sands are evenly distributed throughout the charge, this does not occur in filling through a vat full of solution. This method of filling, moreover, gives a more porous charge, the average weight of a cubic foot of sands as filled into the vats being but ninety-two pounds, figured from the last eight months run; the specific gravity of the original ore averaging 2.7.

When the vat is filled it is leveled off and battery solution is run on for an average of ten days; this battery solution contains a small amount of slimes which form a coating on top of the charge requiring an occasional light raking over to keep a satisfactory leaching rate. The battery solution is followed by barren solution treatment for about six days more when the vat is allowed to drain and a wash water of fifteen tons is put through; the sand is now ready for sluicing, requiring from 100 to 150 tons of water for this purpose.

An average of a large number of sand vats gives 900 tons battery solution and 450 tons of barren solution for one sand vat treatment, exclusive of solution filtering through charge while filling which amounts to approximately 700 tons. This large amount of solution (being nearly ten tons to one ton of sand) together with a total treatment time of about sixteen days has been found necessary, experiments showing that a large volume of weak solution kept leaching through the charge as quickly as possible is essential to a satisfactory extraction.

SLIME TREATMENT.

The overflow from the cones flows to two loading vats which are filled alternately.

There are eight slime vats (including the two for loading) twenty-four feet in diameter and twelve feet deep all of which are connected to two No. 4 centrifugal pumps which can deliver to any one of the slime vats. The centrifugal pumps are four-inch suction and four-inch discharge running at 550 revolutions per minute and handle fifty tons of wet pulp per hour. Quite a large amount of experimenting has been done on the transferring and agitation of the slimes by compressed air replacing the work done by the centrifugals, but these experiments are not yet sufficiently advanced to be of value.

The loading vats are provided with a partition through the center of the vat going down to within thirty inches of the bottom, thus allowing the slimes to settle sufficiently so that the clear solution may be decanted from one side of the partition while the vat is being filled from the other side. The time of filling one of the vats is twelve hours, when the stream is turned into the other loading vat.

The slimes going into loading vat have twelve tons of solution to each ton of dry slimes, the capacity of the vat being 150 tons of solution, it follows that during a loading of twelve hours 150 tons or half of the incoming solution has been decanted off.

This loading vat just filled is decanted as close as possible and transferred to vat No. 1 by the centrifugal pumps, barren solution being added at the same time. The second loading vat, after decantation, is transferred with barren solution to vat No. 2, and upon the decantation of these two vats Nos. 1 and 2, the contents are combined and pumped to a third vat; two more transfers and dilutions are given with barren solution and finally one with water. After each transfer and dilution several hours of agitation are given by pumping out of the bottom of the vat and discharging into the top of the same vat.

It will be noted that the two largest dilutions are obtained on half charges when the contained solutions in the slimes are the richest. A charge of fifty-five tons of dry slimes from a twenty-four hours run gets the following dilutions:

First half charge, 22½ tons, 1 dilution of 85 tons.
Second half charge, 22½ tons, 1 dilution of 85 tons.
Full charge, 55 tons, 3 dilutions of 55 tons.

Making a total dilution of 335 tons or a little more than six tons of barren solution to one ton of dry slimes, the actual figures for the last six months being:

6.38 tons barren solution per dry ton slimes.
0.96 tons wash water per dry ton of slimes.

Theoretical calculations on this amount of dilution, on the assumption that the extraction has all taken place before the first decantation, and taking the value of the barren solution at ten cents per ton, the dissolved gold going out with the slime tails should assay from twelve to twenty cents per ton solution, starting with a head slime solution of $1.00 to $2.00, however, the extraction has not all taken place, but continues slowly throughout the entire treatment, and the solutions finally going out with the slime tails for the last six months of 1903, showed an average value of 46.1 cents per ton, with an average value of a little less than $2.00 for the head slime solution and a value of 10.6 cents per ton of the barren solution.

The decantations are brought down to a pulp containing from fifty-five to sixty per cent moisture. After the decantation of the wash water the top layer of thinner slimes is drawn off and thrown back to the charge following, in this way we obtain a dryer slime going to waste, averaging for the last six months of 1903, 46.7 per cent moisture.

PRECIPITATION.

Only the richer solutions from the sand vats go to the gold tank for precipitation, the balance of sand vat solutions and all the decantations from the slimes going to the battery solution sump.

All standardizing of solutions is done in the gold tank thus getting the benefit of the higher strength solution to assist in the precipitation, this accounts for the higher strength of the barren solution noted before.

Four iron zinc boxes are used, having eight compartments with a capacity of seven cubic feet for each compartment,* hand cut zinc only is used as it offers a better precipitating medium than the machine cut; it is true however that the zinc consumption is heavier with the hand cut zinc. Data for the last six months of 1903 on zinc box flow is given below:

	Tons Solution	Cub. Feet Content Zinc	Tons Sol. per Day	Tons Sol. per Cub. Foot Zinc
July	10,265	194	331	1.71
August	15,130	188	488	2.60
September	14,130	200	485	2.43
October	14,777	212	477	2.25
November	12,793	200	426	2.13
December	12,501	230	403	1.75
Six months 1903		204	435	2.13

Having the measurements of the zinc box flow and the assays on the head and tail solutions we are enabled to keep very close check on extractions; the actual bullion returns exceed the precipitation as shown by the boxes by 3.4 per cent average.

Of the barren solution about one-fourth goes for sand and the balance for slime treatment.

The zinc consumption for the last six months of 1903 is given below:

	Lbs. per Ton Ore Treated	Lbs. per Ton Sol. of Zinc Box Flow
July	1.86	0.310
August	1.08	0.291
September	1.50	0.316
October	1.15	0.267
November	1.37	0.317
December	1.19	0.300
Six months 1903	1.33	0.298

BATTERY SOLUTION.

This solution, of which 1,100 tons per day are used, is pumped by a Prescott 10x7x12-inch duplex pump to stock tank at top of mill, and is distributed approximately as follows:

	tons
To battery	500
To cones	150
To launder	100
To sand vats	350

CLEANING UP.

Clean ups are made twice a month, and are acid treated in the ordinary manner, washed and filtered and taken to the melting room, situated at

---

*Until a short time ago there were in addition eleven zinc barrels to assist in the precipitation, but these are now replaced by another eight-compartment iron zinc box of seventy-six cubic feet total capacity.

some distance from the mill proper, where they are dried in a muffle furnace and fluxed as follows:

10 parts product.
4 parts bicarbonate soda.
1 part borax.
1½ parts sand tails (60% available SiO2).
¼ part iron scrap.
⅛ part flour spar.

This flux has given very good satisfaction, the charge melting easily and quitely and giving a clean and liquid slag. The melting is done in a No. 200 crucible and forced draft is used.

The bullion goes to the U. S. Assay Office at Deadwood, and the slags are shipped to Denver smelters.

Just below the acid tank is located a waste sump having a capacity of twenty-five tons water, into which all of the solutions from the acid tank and vacuum filter are allowed to flow; this solution after the first clean up assayed $12.00 per ton in soluble gold, but now rarely assays over $2.00 per ton. This solution is treated with fine zinc, obtained from the zinc lathe, and sulphuric acid, is well stirred and allowed to settle after action has ceased, when it is found that about ninety per cent of the soluble gold has been precipitated. The sweepings around the zinc boxes after a clean up are also thrown into this waste sump. This waste sump settlings together with the accumulations of mattes is taken care of in a general clean up made twice a year. The mattes are melted down with scrap iron, sand and a small amount of flux, giving a bullion high in copper and carrying about eighty parts gold and 600 parts silver.

All sweepings from melting room and all ashes that happen to assay rich enough are crushed and mixed with the slag shipments.

CYANIDE CONSUMPTION.

The cyanide consumption per ton of ore treated for the last eight months is given below both for the chemical and mechanical losses; the mechanical loss being that going out in waste solutions. The lime consumption is also appended:

	Chemical KCy Cons. lbs.	Mechanical KCy Cons. lbs.	Lime Cons. lbs.
July	0.93	0.67	7.18
August	1.03	0.42	7.91
September	1.07	0.46	7.36
October	1.08	0.38	6.14
November	0.82	0.62	5.48
December	0.89	0.41	5.36
Six months 1903	0.98	0.49	6.54
January, 1904	0.52	0.51	5.03
February, 1904	0.80	0.39	5.95

EXTRACTION RESULTS.

While we are not obtaining a high extraction on our ores, we have the satisfaction of having made a fairly regular improvement from month to month. Cyanide tests have been made regularly on all the ore going to the mill for the past four months; the bullion returns exceed the test extraction by 6.7 per cent.

The following is a compilation of extractions from the starting of the mill, bullion and slag returns being quoted separately:

	Gold% in Bullion	Gold% in Slag	Total Gold%	Total Silver%
First half 1903	46.75	1.81	48.56	22.6
July	50.23	1.39	51.62	26.8
August	58.09	0.49	58.58	26.6
September	60.26	0.80	61.06	18.6
October	63.44	0.49	63.93	17.8
November	67.32	2.02	69.34	21.4
December.	68.23	1.64	69.87	49.2
Second half 1903	61.15	1.07	62.22	26.1
January and February, 1904 . . .	67.66	1.88	69.54	52.2

Of this extraction fifty-three per cent is obtained in the battery, twenty-four per cent from the sand leaching, and twenty-three per cent from the slime treatment as figured from the results of the last four months.*

COST OF TREATMENT.

The working costs of a plant of this nature naturally vary quite a little from month to month, as all expenses incurred are taken up at once and charged out, and where these expenses cover several months or more they cause that particular month to have an unduly high cost sheet; an average of six months run is given whereby the larger part of this discrepency is eliminated:

CRUSHER	Power.	$0.018	
	Labor	0.064	
	Repairs, etc.	0.010	
			$0.092
SAMPLING.	Power.	0.007	
	Labor	0.012	
	Repairs, etc.	0.002	
			0.021
STAMPS.	Power.	0.296	
	Labor	0.108	
	Repairs, etc.	0.058	
	Sand pumps.	0.026	
			0.488

*Since the compilation of these figures considerable improvements have been made, bringing the average gold extraction for the first five months of 1904 to seventy-three and one-half per cent.

LEACHING...............	Pumping solution.............	0.051
	Sluicing sands...............	0.007
	Handling slimes..........	0.019
	Labor......................	0.069
	Repairs, etc.	0.013
		0.159
CHEMICALS.............	Lime.......	0.034
	Cyanide...................	0.368
		0.402
PRECIPITATION	Labor.....................	0.019
	Zinc.....................	0.091
	Repairs, etc...............	0.003
		0.113
CLEAN UP...............	Labor.....................	0.005
	Acid.....................	0.009
	Fuel.....................	0.005
	Fluxes...................	0.005
	Crucibles.................	0.006
	Repairs, etc...............	0.004
		0.034

Assaying...	0.044
Mill Engineers...	0.079
Electric Light...	0.035
General Expense...	0.322

Total Expense of Milling one ton of ore.................... $1.789

The clean up cost figures very close to 8.5 cents per ounce of fine bullion produced, including the cost of marketing the bullion.

In the above costs on power, the fuel, labor and necessary repairs to the generation of steam are included, and the division of the power is based on indicator cards taken from the mill engine, the average cost of a horse power day being 35 cents.

The costs given include every item of expense connected with the running of the plant and takes into account all renewals and changes made, but does not include depreciation of property.

GENERAL.

From a large number of experiments made on the oxidized ores the conclusions reached are that fine crushing, long time treatment, and a large volume of low strength solution give the best results.

Roasting does not materially improve the extraction, neither does the use of oxidizing agents such as bromine appear to help.

Free gold has only rarely been encountered in small quantities and amalgamation tests give no encouragement. Concentration tests show that there is very little concentratable material in the ore and the concentrates obtained are not of sufficient grade to treat further even if the percentage of extraction would justify such treatment.

Chlorination on raw ores gives very low results while on roasted ores no better results are obtained than by cyaniding.

Tests made on the blue ores have given absolutely no encouragement unless first roasted dead at a comparatively fine mesh after which they yield from seventy to eighty per cent extraction by either cyaniding or chlorination.

In our experiments on the raw treatment of blue ores we have been unable to get gold into solution, the maximum extraction by cyanide treatment being thirty per cent with even lower extractions by other methods.

CYANIDE PRACTICE AT THE IMPERIAL MILL AND SOME COMPARISONS OF DRY WITH WET CRUSHING IN CYANIDE SOLUTION.

By John T. Milliken, E. M., Superintendent Imperial Mill.

[Paper read before Black Hills Mining Men's Association, August, 1904.]

The object of this paper is to describe briefly the reduction of ores by the "dry fine crushing" process, in an up-to-date dry crushing mill. and at the same time make a few impartial and honest comparisons with an up-to-date and modern "fine wet crushing mill."

There are but two cyanide mills in the Black Hills reducing ores by the dry fine crushing process—the Golden Reward Mining Company's mill, with a monthly capacity of 5,000 tons, and the Imperial Gold Mining & Milling Company's mill, with an average capacity of 4,000 tons per month. The excellent results attained by the Golden Reward Company in their dry crushing cyanide mill need no mentioning here. so the writer will confine himself to the mill of the Imperial Gold Mining & Milling Company, as it is a more modern mill, and equipped with the necessary machinery for economic handling of the ore.

THE ORE.

The ore is from the Potsdam, extremely hard, tough, close grained, highly siliceous and the gold minutely disseminated. Eighty per cent of the ore as it comes from the mine is fairly well oxidized. The remaining twenty per cent is what is known as the blue ore. This blue ore is very refractory, close grained, carries from four to five per cent iron pyrite. traces of arsenic, as arsenopyrite and antimony. In fact it is this blue ore that makes it difficult to recover seventy-five to eighty per cent from the low grade gold ores of the Black Hills, but as the blue and oxidized ores are mined together, it is absolutely impossible to eliminate all the unoxidized ore.

MILLING.

The Imperial mill is a flat site mill, each department having its own building independent from the mill proper. This form of construction affords perfect control of the dust and confines it to the building wherein it is produced; which, on the other hand. or in a hill-side mill, where all departments are practically under one roof. dust produced in any one portion flies through and permeates the whole mill. The engine room, the leaching and precipitating department, are absolutely separate from the fine crushing department, which is the only department where dust is produced.

The mill is well provided with railroad spurs, both on the Burlington & Missouri River and Chicago & Northwestern Railroads. The ore is received from the mines in bottom dump cars. and immediately dumped into railroad bins. and is drawn directly from here into a 10x20 Blake crusher. running 250 revolutions per minute. and is here reduced to rock varying from fine sand to pieces having an extreme dimension of two inches. The discharge from the crusher feeds directly into a twelve-inch bucket elevator, fourteen-inch nine-ply belt speed 300 feet per minute. This elevator discharges directly into a 3x6-foot Hexagon screen revolving fifteen revolutions per minute, which gives a sized product varying from fines to ore having an extreme dimension of one inch. The oversize from this screen returns to a pair of 16x36 Davis rolls, speed sixty revolutions per minute, and the undersized is carried by a sixteen-inch belt conveyor to storage bins having a combined capacity of one thousand tons. The discharge from this 16x36-inch rolls joins the feed from the crusher and is again elevated to the sizing screens, so nothing goes to storage bins but carefully sized material.

This is known as the sampling department, or preliminary crushing. The sample is cut from the sized ore as it leaves the above mentioned screen, and is cut out with a Vezin Automatic Sampler. This sampler cuts out accurately one-twenty-fifth of the ore which is spouted to steel covered floor provided for it, and is here finally reduced by splitting with a Jones sampler to a definite weight, and run through a pair of 12x12 sampling rolls, cut or reduced further with a riffle, thoroughly dried, and finally ground in a No. 2 sample grinder to about eighty mesh, thoroughly mixed, cut with a riffle to about ten ounces and bucked through 120 mesh screens, and two samples made.

This amount of work necessary to obtain a representative sample of a particular lot of ore, may seem unnecessary to some, but the value to a plant of trustworthy sampling far exceeds the money cost, and figures based on any other but an accurate head sample are very misleading.

FINE CRUSHING.

The ore from the above mentioned bins is now fed onto a twelve-inch belt conveyor by an automatic fine ore feeder, this feeder provides a steady and unvarying feed to the rolls, an absolute necessity for good work. The belt conveyor discharges into a ten-inch bucket elevator eleven-inch seven-ply belt, speed 300 feet per minute, which discharges directly into a No. 2-4 Tube Argall Dryer. This dryer consists of four fire brick lined steel tubes, nested together inside of two tires, and provided at the ends with heads which serve to feed and receive the wet ore, and to discharge the dried ore. These tubes are twenty-five inches in diameter and twenty-five feet long. This dryer has given excellent satisfaction, requiring a minimum of fuel, one horse power for revolving under full load, and the repairs are nominal. It is placed on a slight incline three-fourths of an inch to the foot and makes two revolutions per minute.

The temperature maintained in this dryer at the Imperial, is close to 300 degrees Fahrenheit, this temperature dehydrates the ore thoroughly, and very materially increases the filtering or percolation in the cyanide vats. Figures will be given further on relative to fuel, repairs and condition of ore after drying.

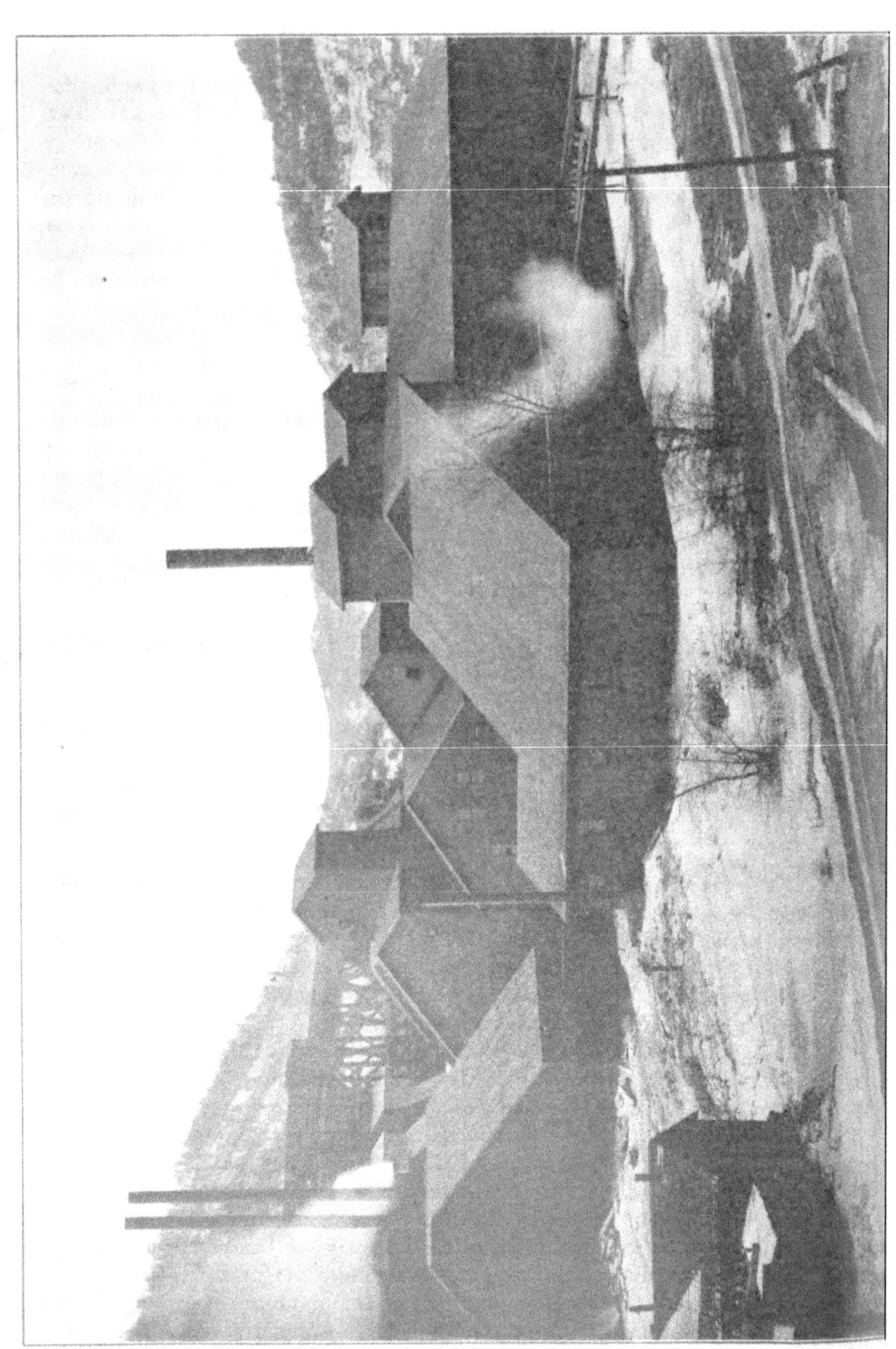

The dryer discharges directly into a roller chain continuous bucket elevator, provided with twenty-five-inch traction wheels for head and boot pulleys, and has a speed of 150 feet per minute.

This chain elevator discharges directly into a 2½x6-foot cylindrical screen, revolving fifteen revolutions per minute, and covered with five mesh No. 16 wire, opening .137 inches. The oversize discharges and passes to a pair of 16x36 Davis rolls, running fifty-five revolutions per minute, the discharge from this roll joins the feed from the dryer and is again elevated and sized. The undersize from the above screen passes to the finishing rolls, that is, one-half of the feed from the roughing rolls passes to one of the ten-inch bucket elevators, twelve-inch seven-ply belt, and is elevated and discharges into two cylindrical revolving screens four feet in diameter, eight feet long, revolving fifteen times per minute, and covered with sixteen mesh No. 21 steel wire cloth, size of opening .0305 (finished pulp). This system gives the ore from the roughing roll a thorough screening before it passes to the finishing rolls, 16x36 Davis rolls, running seventy revolutions per minute. This is one unit. These screens are completely housed in with steel housing, and are practically dust proof. The undersized or finished pulp passes to a ten-inch bucket elevator, twelve-inch seven-ply belt, and is immediately elevated to finished ore bin. The oversize from the above screen discharges directly to the finishing rolls, recrushed and joins product coming from roughing roll, and is elevated again to finishing screens. There are two finishing rolls, each receiving one-half of the feed from roughing roll, each finisher having two screens four feet in diameter, eight feet long, one screen on each side of the ten-inch bucket elevator. This close sizing from crusher onto finishing rolls has been proven absolutely necessary, owing to the extreme hardness of the ore, but the rub lies in the toughness and density of the ore, which from the finishing screens since February 1st, have averaged 100.7 pounds per cubic foot, and the general average for twenty months has been ninety-eight pounds per cubic foot. Since this close sizing has been inaugurated, all the rolls, more especially the finishers, have been running quietly, without shock or undue strain, and has cut the steel shell cost from eight cents per ton to four and three-fourths cents, and reduced short shut-downs to a minimum.

CYANIDE TREATMENT.

The ore from finished pulp bin is fed directly on to a nine-inch screw conveyor, eight feet long, and at this point the ore is sprayed with the cyanide solution in order to do away with the dust in the leaching room, which it does very effectually. This screw conveyor disharges onto a fourteen-inch belt conveyor, which in turn discharges onto a fourteen-inch belt elevator, fourteen-inch nine-ply belt, speed 300 feet per minute. The discharge from the elevator feeds to another fourteen-inch belt conveyor, passing over the leaching vats, and from this conveyor all the vats are filled. There are four leaching vats, six feet deep, thirty-five feet in diameter and built of three-sixteens-inch steel, and when filled within three inches of the top, hold approximately 275 tons of ore. In filling the vats the ore is always charged into standardized solution, five pounds cyanide to a ton of water, so when vat is filled it is practically saturated. However, after vats are filled and leveled off, the surface is flooded with standard solution and charge allowed to stand until perfectly quiet, or thoroughly saturated. Leaching is then started and continuous percolation kept up until charge is washed. Time required to fill a vat is fifty hours, and treatment will average eight days. The amount of cyanide solution in tons required for treatment, taking 100 charges as an average, is as follows:

65 tons 5 pounds solution for saturation.

70 tons 3¼ to 4 pounds solution returned from strong sump.

20 tons of 2½ to 3 pounds solution returned from weak sump.

50 to 55 tons of wash water.

Practically six-tenths of a ton of solution to each ton of ore treated. Larger amounts of the solution have been often used, as much as 200 tons of weak and strong have been returned to a vat without any apparent increase in extraction, and amounts as low as 600 pounds of solution to one ton of ore have given the same tailings as when the normal 1,200 pounds per ton of ore has been passed through the charge. The maximum value of the effluent solution is reached in about fifty to sixty hours after the vat has begun leaching, and has often run as high as two ounces per ton, and a very fair average for all charges would be about $30.00, or one and one-half ounces per ton. The dissolution seems to take place very rapidly, and when the above value is reached it begins to decrease very rapidly, but gradually. All the solutions coming from leaching vats, including the wash water, are passed through zinc boxes.

The vats are provided with the usual filter grating, covered with coco matting and eight-ounce duck, and have five eight-inch diameter Argull sluicing gates, one in the center, and one in each quadrant, and are sluiced out in the usual manner with hydraulic pressure.

PRECIPITATION.

In the precipitation department there are two gold storage tanks four feet deep and twenty feet in diameter, built of three-sixteens-inch steel. These tanks are for the effluent and weak solution. There are two sumps immediately under the zinc boxes, and are the same size as the gold storage. There are two double compartment zinc boxes, fourteen compartments two feet square and two feet deep to each box, or twenty-eight in all. These compartments hold when closely packed fifty-five pounds lathe turned zinc.

The present zinc area required for precipitating the values from 130 tons of ore per day is 104 cubic feet, containing 800 to 1,100 pounds zinc shavings, that is just before a clean up takes place, the boxes will contain 1,100 pounds zinc shavings, the first two compartments will contain practically 150 pounds each, and it may sound incredible but it is a fact that fully 95.5 per cent of the precipitation takes place in the first two boxes on the strong side.

There are only three boxes of the fourteen on the weak side used, three of them being sufficient to keep sump thoroughly clean. The solutions are standardized by pumping from sumps in to two standardized tanks, sixteen feet deep and fourteen feet in diameter, and are also built of three-sixteenths-inch steel.

EXTRACTION.

In the mill there is no separation made, of the sands and slimes. The ore is treated as it comes from the rolls, sands and slimes together, a system which permits of the closest watch possible on extraction. The amount of gold solution passing through zinc boxes is readily measured, constantly sampled and checked with the total solution, returned from sumps, and with the drip samples taken on each tank continuously during its entire leaching period, including washing. This extraction compared with the extraction figured on the difference between control and tailing assays, is about 1.4 per cent higher. The actual bullion return checks with the solution extraction within 1.5 per cent, the actual bullion return being 1.5 per cent higher. The actual bullion return since mill was started is 78.2 per cent.

The operating cost per ton of ore treated, for the last running twelve months, based on a daily tonnage of 100 tons per day is as follows:

LABOR AND SUPPLIES.

	Per ton
Labor	$0.580
Fuel for power	0.255
Fuel for drying	0.110
Lubricants	0.051
Roll steel	0.057
Equipment supplies	0.045
Cyanide	0.120
Zinc	0.070
Melting supplies	0.014
Assay supplies	0.015
Miscellaneous supplies	0.040
Insurance	0.030
	$1.387

The tonnage now being maintained is 130 tons per day of twenty-four hours. And when contemplated improvements are perfected, the operating cost will be further reduced.

THE CLEAN UP.

The usual sulphuric acid method of cleaning up is used. And owing to the small amount of sol passed through boxes and its freedom from slimes, the precipitates are high grade and in an excellent state for dissolution with the acid. Considerable coarse, or zinc that has begun to decompose, is also removed, with the fine precipitates.

The precipitates are washed free from cyanide solution by being discharged directly from the boxes into a vacuum filter two feet deep and four

feet in diameter, provided with a filter bottom of coco matting, ten-ounce duck, and this covered with a heavy quality of lawnsdale muslin, which reaches up over the top of the filter proper. The precipitates are removed from here to a steel lead lined tank two and a half feet deep, five feet in diameter. Hot water is now added so as to give a final acid dilution of one in six, the precipitates having been weighed as they were transferred to the dissolving tank. Acid is now added gradually until it reaches the above ratio, and when this has been reached the zinc precipitates is well dissolved, no coarse zinc being present.

The dissolving tank is now covered and the contents given a vigorous boil for one hour. The boiling is now stopped, the sides of the tank now carefully washed down with a small stream of hot water, and the contents run out into another vacuum filter, the same size as the one described, and the liquid drawn off slowly but steadily, until dry, three hot water washings are now given, precipitates being carefully stirred each time until they are all in suspension. The precipitates are then dried with the vacuum until they begin to crack. They are now removed and placed in sheet steel pans, (the bottom of these pans are first covered with paper, this keeps the slimes from sticking to pans), and dried slowly in cast iron muffles, the flame passing over the top, 390 pounds of precipitates as they come from the boxes have been cut down, thoroughly washed and placed in the dryer in eight hours, care and cleanliness not being sacrificed for speed.

The precipitates after drying are fluxed with twenty-two per cent borax glass, twenty per cent soda, fifteen per cent silica, covered with crude borax and fused in wind furnaces. No additions are made during fusion.

The bullion has averaged for the past six months 935.7 per cent. fine metal.

Average time spent on clean up eighteen hours. The slimes are dried during the night, melting started in the morning, two men carry out the work. Average cost per ounce of bullion recovered 7.1 cents. The bullion for the past four months has averaged as follows:

504.5 per cent silver
431.2 per cent gold
———
935.7

The slags, pot scapings and mattes are run to base bullion with litharge.

COMPARISONS.

The following are a few comparisons with the fine wet crushing plants, and some of the so-called bad features of the dry crushing discussed.

One of the drawbacks to the fine dry crushing mill is the dust It is indeed the principal one, but, however, is very much magnified as to its numerous drawbacks by popular talk. The dust is hard on the pulmonary organs of the workmen, and disagreeable to work in, and hard on bearings and machinery. But with the present improvements for handling dust, and the mill designed with a view to keeping the dust confined to that particular department where it is produced, a fine dry crushing plant will not be a bad place to work, and there will be no difficulty in keeping men.

Another point put up against the dust or fine dry crushing is the loss in values from the escaping dust. There is a loss in values by escaping dust, so is there a loss also in dissolved gold in the wet crushing process, and a loss that far exceeds anything the writer has ever seen in good dry crushing plants.

I will describe briefly the method of handling the dust produced at the Imperial mill, and give a few figures on the same.

The only dusty and dust-producing portion of the mill is the fine roll room. All the screens are steel housed, elevators are thoroughly housed, and heads are covered with ten-ounce duck. All these housings and places where dust is generated are connected up with four-inch galvanized iron pipe, all these pipes are connected to one central head attached to two thirty-five-inch exhaust fans, that draw and collect the dust and blow it into two No. 39 Prinz Rau dust collectors, which accumulate the dust and discharge it automatically, into small boxes attached to wheel barrows, and it is mixed with the finished pulp and treated with it. The amount of dust collected by these two dust collectors from 130 tons of ore every twenty-four hours is 3,222 pounds, 97.2 per cent of it passes through a 150 mesh laboratory screen, assays $10.65 per ton and was drawn from ore running $8.00 per ton. This dust weighs forty-nine and a half pounds per cubic foot. The original ore weighs 100.7 pounds per cubic foot.

The dust produced by the dryer is caught in a dust flue connected with it. This dust flue is 4x5 feet high in the clear, and is 107 feet long. The dust produced by dryer amounts to three-tenths of one per cent, and assays twenty-five per cent less than the original heads, or the average of the dust since mill started $5.00 per ton, a very unusual state of affairs for dust values, for as a rule the dust will run from one and one-half to three times the value of the original ore.

The dust that accumulates throughout the mill hardly ever exceeds the value of the ore.

The dust that escapes from the buildings is lost. But inferring from dust saved by dust collector, tank weights and bullion recovery, based on accurately sampled ore, it is undoubtedly below one-half of one per cent.

Comparing this dust proposition with the slimes loss in the wet crushing mills, there seems to be food for thought. I will base all comparison on results and information from Mr. Gross, Superintendent of the Penobscot mill, as the mill is one of the most up-to-date and modern wet crushing mills in the Black Hills, and great care being given to arrive at and determine accurate results.

Percentage of slimes produced in crushing wet with stamps at the Penobscot mill is 51.8 per cent of the total ore crushed. These slimes are conducted to tanks with thirteen tons of cyanide solution to each ton of dry slimes, and are here agitated and washed by decantation. They are given three dilutions with weak cyanide solution decanted each time but with several hours agitation with each decantation. They are finally given one water wash, and then sluiced out, containing 46.7 per cent moisture, and 46.1 cents per ton dissolved gold, the amount of undissolved gold I have no figure on.

Forty-six and one-tenth cents per ton dissolved gold escaping would be at the rate of twenty-four cents, figuring 51.8 per cent of the ore treated as slimes, or three per cent on an ore running $8.00, a very high dust loss indeed. The Dakota M. and M. Co.'s loss is also three per cent.

Another point might be mentioned regarding moisture—they say, why dry your ore, it will take up and use so much of your water. Well, it does take out with it some water, an average on 100 vats shows the pulp contains on going out 12.0 per cent moisture. This is several per cent higher than necessary if water were short. However, this is considerably less than 46.7 per cent moisture.

Drying the ore is one expense the wet crushing mills avoid, but the disadvantage of having to dry your ore is more than offset when it comes to leaching the pulp. It is a well known fact that when ore is heated to a temperature of 300 degrees Fahrenheit, it is dehydrated and the leaching accelerated five or six fold. At the Imperial mill tests have verified this.

The following pulp will leach at the rate of thirty-five tons per twenty-four hours, when thoroughly dried—vat six feet deep, containing five feet seven inches of ore.

		Per cent
Remaining on	20 mesh	2.5
Remaining on	30 mesh	11.0
Remaining on	40 mesh	10.0
Remaining on	60 mesh	18.0
Remaining on	80 mesh	9.0
Remaining on	100 mesh	4.0
Passing 100 mesh		45.5

Twenty-one and one-half per cent of this 45.5 per cent will pass a 150 mesh screen. These screens are the usual brass wire screens of the laboratory.

Cost of drying ore at the Imperial mill and maintaining dryer in good repair, is:

	Per ton
Fuel	$0.110
Repairs	0.010
Labor	0.0175
Power	0.0030
	$0.1405

Drying at the Golden Reward Mining Company mill costs:

	Per ton
Fuel	$0.0700
Labor	0.0330
Repairs	0.0100
Power	0.0030
	$0.1160

Another point in favor of drying, and a good one, is in place of the cyanide solution coming in contact with frozen ore, which is the case in a wet crushing plant during the winter months, it comes in contact with warm ore. This insures an even temperature throughout the year, and consequently a uniform extraction.

ELEVATORS.

Elevators and the power to elevate the ore so much is another thing the wet crushing mills claim they are free of. They are free of elevators but not free from the cost of elevating. In almost all ore milling operations one of the most useful and necessary, and at the same time most objurgated pieces of apparatus is the belt elevator. Yet, if a belt elevator is properly designed it should not give any more trouble than more complicated machinery, which is exposed to equally hard conditions of service.

All the elevators in the Imperial mill are provided with the best rubber belts and malleable iron buckets, have a speed of 300 feet per minute, strict attention being paid to the feed and discharge.

There are six rubber belt elevators in the Imperial mill and one continuous bucket chain elevator.

No. 1 elevator. . . . . .42 feet center to center.
No. 2 elevator. . . . . .22 feet center to center.
No. 3 elevator. . . . . .26 feet center to center.
Nos. 4 and 5 elevators . . . . .38 feet center to center.
No. 6 elevator. . . . . .50 feet center to center.
No. 7 elevator. . . . . .35 feet center to center.

Elevators Nos. 4 and 5 are for the finishing rolls.

No. 1 elevator handles 130 tons in 12 hours.
No. 2 elevator handles 130 tons in 24 hours.
No. 3 elevator handles 130 tons in 24 hours.
Nos. 4 and 5 elevators handle 65 tons each in 24 hours.
No. 6 elevator handles 130 tons in 24 hours.
No. 7 elevator handles 275 tons in 52 hours.

The above seven elevators require 1.90 theoretical horse power and allowing an equal amount for friction, the total is 3.8 horse power. There are also four belt conveyors, a liberal allowance for power in this case is four horse power, making a total power for elevating and conveying ore 7.6 horse power.

The cost of operating is:

	Per ton
Power.	$0.016
Repairs.	0.018
	$0.034

or three and four-tenths cents per ton of ore treated.

The following are Mr. Gross' figures on elevating at the Penobscot mill:

Running and maintaining of two Fiernier sand pumps for the past year. Lift from battery discharge to classifiers 20.5 feet 36,200 tons pulp and 1,810,000 tons solution.

Power. . . . . 2.45 cents per ton of ore.
Supplies. . . . . 0.03 cents per ton of ore.

2.48 cents per ton of ore.

Repairs on two Fiernier Sand pumps for the last year:

Material.......... 0.21 cents per ton.
Labor............. 0.06 cents per ton.
                   ————
                   0.27 cents per ton.

Or 2.75 cents per ton of ore for elevating from battery discharge to cone classifiers.

The cost of handling slime during washing and decantation is:

Repairs on two No. 4 centrifugal pumps for one year, handling 220,000 tons of dry slimes, with approximately 1,500,000 tons solution, (this includes transferring and agitation) lift of pump twenty-nine feet under average head of eighteen feet, average per lift eleven feet, only repairs being a new shaft for each pump.

Material.......... 0.04 cents per ton of ore in mill.
Labor............. 0.05 cents per ton of ore in mill.
                   ————
                   0.09 or 0.015 cents per ton of ore handled.
Power............. 2.02 cents per ton of ore in mill.
Supplies.......... 0.09 cents per ton of ore in mill.
                   ————
                   2.11 or 0.35 cents per ton of slimes handled.

The above figures based on the ore handled in the mill show a cost of four and three quarters cents per ton of ore handled.

There is also in the neighborhood of 1,100 tons of solution per day pumped from the sump to battery storage tank, with an average lift of eighty-five feet, a reasonable allowance for this would be about six horse power, or the cost would be approximately for power, labor and repairs, about two cents per ton, or a total of six and three-fourths cents per ton for elevation of material, as against $0.034 per ton in the Imperial mill. The Dakota M. and M. Co.'s cost for elevating is .06.85 cents per ton.

The following are figures given by the Golden Reward Mining Company, and are based on a tonnage of 101,682 tons in a dry crushing plant:

	Per ton
Roll steel.	$0.0426
Bab belt metal.	0.0182
Elevator buckets	0.0016
Elevator belts.	0.0044
	————
	$0.0668

Or 6.68 per ton.

The Golden Reward mill has but four elevators and no conveyor.

The cost of steel for crushing seems to be very much the same thing for the wet and dry plants.

Comparison on consumption of chemicals, figures based on 100 tons per day in each mill, the ores in each case being very similar as to general character and value.

PENOBSCOT WET CRUSHING PLANT.

Average strength solution, two pounds; cyanide chemical loss, .98 pounds; cyanide mechanical loss, .49 pounds; pounds of zinc in contact, 1,800 to 2,000 pounds,; zinc consumption 1.33 pounds.

Total cyanide consumed 1.47 pounds per ton. Tons of solution passed through zinc per ton of ore 4.65 tons.

IMPERIAL DRY CRUSHING PLANT.

Average strength solution, 4.5 pounds; chemical and mechanical loss, cyanide, .50 pounds; pounds zinc in contact, 800 to 1,200 pounds; zinc consumption, .70 pounds.

Total cyanide consumed .5 pounds per ton.

Tons of solution passed through zinc per ton of ore .8 of a ton including wash water.

The average amount of solution passed through the sands or coarse portion of the ore in the wet crushing plants is practically ten tons to every ton of ore. This is due principally to the fact that it is absolutely necessary to keep the strength of working solution as low in value and cyanide as possible. A good average for working strength in cyanide is one and one-half pounds of cyanide to the ton of water, such strong solutions as are used in the dry crushing plants are prohibitive, owing to the large mechanical losses it would entail. On the other hand the value of the solution coming from the batteries must be kept down as low as possible, otherwise the loss in dissolved gold escaping with the slimes would be ruinous..

CONCLUSION.

In construction the wet crushing mill is much the simpler, having fewer pulleys, belts, generally only one elevator, and no screens except the battery screens.

The cost of construction per ton of ore, will be practically the same in mills designed to treat similar ores.

The wet crushing plants are more cleanly and easily kept so. On the other hand, in the scope or elasticity of the processes, the wet crushing is limited in many ways. First, on an ore running high in values and producing a very large percentage of slimes, and where these slimes would carry values exceeding the original or heads by 150 to 200 per cent, not an unusual condition of affairs. One very unusual characteristic of gold ores, and one that makes the low grade gold ores of the Black Hills particularly applicable to wet crushing, is the fact that the slimes carry less values than the sands or ore from which they were separated. Second, On ninety-five per cent of the low grade sulphide ore requiring roasting, the Black Hills ore for example, the blue or unoxidized ore, is developing very rapidly, and in no distant time the mills will have to resort to roasting and cyaniding, and all the ores the writer has experimented on will have to be crushed to at least twelve mesh, and sixteen and twenty mesh have given the better results.

A representative sample from the McGovern property, gave the following results, on roasting at different meshes.

Sulphur contents of sample, 2.68 per cent S.
Gold contents of sample, $10.80.

These samples were roasted in the muffle furnace, and the temperature kept as near as possible to the usual temperature maintained for low sulphur ores. That is from 500 degrees Fahrenheit to 1,500 degrees Fahrenheit, the finishing heat.

Samples were crushed to—	S. contents before roasting per cent	S. contents after roasting per cent	Extraction per cent
Six mesh No. 16 wire, Size of opening 0.102	2.68	.972	14.2
Eight mesh No. 16 wire, Size of opening .060	2.68	.611	19.75
Ten mesh No. 17 wire, Size of opening .042	2.68	.220	52.1
Twelve mesh No. 18 wire, Size of opening .034	2.68	.161	66.8
Sixteen mesh No. 20 wire Size of opening .0275	2.68	.100	77.9
Twenty mesh No. 23 wire Size of opening .023	2.68	.071	83.7

The time and strength of solution was the same in each case. A six pound solution was used. Cyanide consumption decreased as extraction increased, owing to imperfect roast. The extraction on the sixteen and twenty mesh product took place extremely rapid, and cyanide consumption was .83 of a pound. The pulp leached beautifully and the leaching was increased about eight times over the leaching in the raw state.

Roasting test on the unoxidized ores from the Imperial property in Blacktail, and the Bertha property in Ruby Basin, all show good extraction.

Cynaide roasting tests were made on the blue ore of the Portland Company on Bald Mountain, and it proves to be an excellent roasting proposition.

The above tests point very strongly to fine dry crushing and roasting. The Imperial is designed with a view to roasting their unoxidized ores, and will be put into commission at no distant date.

The unoxidized ores of the Hills, when crushed to about twenty mesh, will yield on an average from twenty-five to thirty per cent of their values. A good average value for the blue ores of the Black Hills is about $11.00, which when treated raw will yield practically $2.50 or $3.00 per ton, but on the other hand if properly roasted will yield at least $9.50 per ton.

This roasting, if done in a thoroughly modern, well-designed plant, equipped with one of the first class straight line furnaces, and proper skill applied, can be done safely for the following figures, based on a tonnage of 100 tons per day.

	Per ton
Fuel	$0.550
Labor	0.065
Repairs	0.100
Power	0.010
	$0.725

Fuel is figured on Sheridan coal at $3.50 per ton at the mill, and the labor item is for one man on each shift watching and firing furnace. It is safe to say that 75 cents would be an outside figure for roasting the ore.

The general conclusion is, that a modern and well designed dry crushing plant will treat the same ores that a wet crushing plant will, and a greater variety of them, and do it cheaper and extract a larger percentage of the values.

CYANIDE PRACTICE IN THE BLACK HILLS.

BY JOHN RANDALL, SUPERINTENDENT OF THE LEXINGTON HILL MINE.

[Paper read before Black Hills Mining Men's Association, June 19th, 1902.]

MR. CHAIRMAN AND GENTLEMEN:

The difficulties which ordinarilly beset the cyanide man make his work somewhat alluring when we remember that leaving out the man who is always looking for trouble, the human race is generally happiest when it has some difficulties to encounter. To begin with, the ore of good value usually goes to the smelter or is shipped to distant parts to undergo highly refined and expensive processes of treatment whereby is recovered nearly or quite ninety-five per cent of its value. The waste from this shipping ore is often denominated cyaniding ore. The cyanide man must be content with treating mud and dirt and leaves and roots and sods, sparingly mixed with rock, and he must get his customary extraction of 104 per cent., more or less, if he would satisfy the demands generally coming from the office for bullion.

The definition of ore is a rather elastic one. For the purpose of the mill man a gold ore may be defined—any material containing enough gold to be milled at a good, round profit, but the more widely accepted definition is far more comprehensive, and includes anything from sea water to mine waste or other waste material probably including the discarded corsets, old shoes and superannuated cats in the back yard of the next lot, and which only a portion of the year is hidden by a charitable mantle of snow from mortal eye and metallurgical investigation. But seriously, the low grade ores of the Black Hills have stimulated improvement which cannot fail to be of incalculable benefit to the district and to the world at large.

Some tribes of Eskimos require every male child at a certain age to pass through the ordeal of being tied into his boat and thrown into the sea. If he rights his craft and comes out he is regarded as being able to make his way in a region where the conditions of bare existence are extremely severe, but if he drowns it is believed he is better thus out of the way. Now the cyanide man who escapes the righteous indignation of the manager who has an insatiable desire for bullion, or the mill man who does not die a merited death in the attempt to discover gold in mine waste, is by low grade ore conditions forced into a position for improving the state of the art wherever possible, and as might be expected, marked improvements have been made right here among us and along all the various important lines of cyanide work.

This mention of "merited death" may require explanation. There will be millions of tons of good low grade ore left in these Hills after the present generation has passed away, and no penalty or punishment is too severe for the man who wastes his time on material too low in value to yield a good, liberal profit above the general expenses of a company and depreciation of plant, besides the usual items charged up to cost of mining and milling.

But this paper will be confined as nearly as possible to a single phase of the subject, namely, crushing in solution, and separate treatment of sands and slimes, as it seems to the writer that herein is the most favorable field for advancement under the particular conditions of this locality.

When we recollect how often we are obliged to go back to first principles, even in these days of marvelous advancement, and note how really little there is new under the sun, we are not surprised that the old, noisy stamp battery is taking its place as a potent factor in this work with its freedom from complicated details and its ability to handle anything that comes from upon, in or under the earth, provided it is fed enough hard rock to keep the shoes from pounding iron, at least a part of the time. Crushing in solution has come to stay although the first attempts here in that direction, like all new things were not fully appreciated.

In the middle of the summer of 1899, a man began fitting up a little old tumble-down stamp mill in Deadwood gulch for the purpose of stamping ore in cyanide solution. He was not altogether a brilliant man, but a hard and conscientious worker, a man of infinite patience, and he had an idea. He was financially backed by a few men of slender means who had a large amount of ore too low in value to be treated by any process yet practiced in this region. People laughed at the attempt to settle and circulate enough solution in a mill to supply stamp batteries. They were told that the experiment was being proven an undoubted success in New Zealand, it having been introduced there about the beginning of that year. Most people are too busy to keep well posted in geography, but we all know that New Zealand is somewhere south of the equator, and the minds of some no doubt reverted to the picture that long ago greeted them from the pages of the primary geography—the picture of the south sea islander, clad in nothing but his modesty and a bunch of dried grass fastened to his loins by a rawhide throng. Then they laughed again, But the rattle of the stamps drowned the laugh, and the stamps kept on rattling. As a matter of fact, the New Zealand practice did not prove anything for Black Hills ores, for there they treated ore that previously yielded a profit with a recovery of only sixty-five per cent by dry crushing and hot pan amalgamation. (Trans. Am. Inst. M. E. September, 1899.) As soon as the old ramshackle mill up the gulch showed that the ore could be treated at a profit some of those interested wanted to put up a big mill, just like the other big mills, a dry mill, of course, and be up to date and in fashion. Now it is a safe and most laudable thing in mill construction to keep as near as possible to the beaten path. This cannot be too strongly impressed upon the mind of every man who has in his hands the disposal of a company's capital. It is better and safer to allow the "other fellow" to do as much of the experimenting as possible. But when a mill man becomes afflicted with an idea, and covered with mud, fashion counts for very little indeed. So the stamps kept on rattling until they won out and the result was the mill of the Dakota Mining and Milling Company now running in

Deadwood and probably capable of treating ore at a less cost than any other fine crushing mill in the world, excepting of course, the plants treating free milling ore. Before these first experiments had proceeded very far, however, the Portland Company leased a stamp mill at Gayville, and put the process in operation with important improvements.

One important characteristic of this process is the separate treatment of the sands and slimes, which on the score of economy presents an apparent contradiction. The carrying out of two materially different processes in the same mill calls for a greater first cost of the plant as well as more detail in its operation, but I am ready to predict that separate treatment is destined to be the rule of practice with the siliceous ores of the Black Hills. There will probably be exceptions to this rule in the form of well conducted mills favored with a uniform supply of ore of special quality, but exceptions often prove a rule. The discussion of the general question of separate treatment opens a large and interesting field of study. A mill should be as free from complexity of details as is consistent with automatism. However automatic operation does not always mean the highest economy. It is easy to overestimate the value of a machine process. There are many persons besides Helen's babies that like to see the wheels go round, and this common weakness of human nature is responsible for an enormous amount of misdirected effort in the building of thousands of complicated machines that cost many times more than they are worth. In following this idea of doing things by machinery, the earlier attempts at cyaniding were along the lines of barrel chlorination somewhat hastening the process, to be sure, but it was soon found that in addition to the high cost and limited capacity of the barrel it cost good money to turn that barrel. Various agitating devices have been tried, but while the political agitator, the spiritual agitator, the social agitator, and many other choice brands of agitators seem to have their field of usefulness, the man who seeks to agitate wet sand, by mechanical means at least, has not so far succeeded in dazzling the world by his exploits. The best profit margin is usually made by the man who allows his sand to run into a big leaching vat, and then leaves it alone. A chemically active solution is actively percolating through the sand by gravity which costs nothing. After depositing its values in the zinc box, this solution oxygenated, aerated and otherwise rejuvenated, spits on its hands, and again, and then again, goes after that sand until we obtain, as everybody knows, the customary extraction of 104 per cent, more or less. During all this time the mill man has been leaving the sand alone, and giving his solution a chance. Oh, how many knotty problems would find a ready solution if we could only give the solution a chance. How many of us would be benefited in mind, body or estate if we had only known when to leave even a good thing alone. It is a little humiliating to note that with all our twentieth century advancement we have been forced to go so far back to first principles in the treatment of wet sand—that we have not materially advanced beyond the ash barrel stage of leaching as probably first practiced by Mother Eve when she made her first batch of soft soap, and soft-soaped Adam into eating that apple that gave him such a thirst for more knowledge and such a desire to "see the wheels go round." But this is where we are at, for the present at least, and as far as sand is concerned, is the conclusion of the whole matter.

To many this declaration may seem rather sweeping, particularly in these days of improvement and discovery, but I feel that it is a safe statement of a general rule. A general truth or a general rule has exceptions. For instance, with separation of slimes and cleaner sand the tendency is toward higher leaching vats and deeper ore charges. Now a good mill man will see that his solution carries plenty of free oxygen before it is put into the sand, but with certain iron compounds in the ore, or for other reasons, there may not be enough oxygen to last down through a deep charge, or any charge with a slow leaching rate. The result may appear in low extraction, particularly near the bottom of the vat. A convenient remedy might be found in the introduction of compressed air at some stage of the process, which is probably as well effected under and through the filter cloth as by a special system of perforated pipes. The consequent slight movement and rearrangement of the ore particles would be a benefit in some cases, in fact a cheap substitute for the South African system of double treatment where the sand is shoveled into a second vat to undergo a second leaching. However, with a clean material thoroughly freed from slimes the apparent advantage of moving the sand might disappear.

But all crushed ore is not sand. "Ay! there's the rub." Many kinds of rock are prone to break into an almost impalpable powder called dust by the ordinary mortal, but universally known as "slimes" among the demons who infest wet crushing mills. Some years ago ores that slimed badly were not generally regarded as amenable to treatment by wet extraction methods. Then came improvements in crushing rolls until they became perfect marvels of the mechanics skill, expensive to buy and more expensive to maintain, but by the gradual reduction method, with a careful arranged system of screening between each break they did the work where other means failed, produced a fairly leachable pulp and scored a substantial advance in metallurgical progress. But some ores, particularly the siliceous ores of the Black Hills, were found to require very fine crushing, and then the old trouble reappeared—too much dust, imperfect leaching, and uncertain recovery of values. Men some time ago learned not to agitate sand on account of the expense, and later after many failures with a thousand patented devices have generally learned not to attempt the leaching of slimes. The only exception that may be taken to this statement seems to be in favor of the filter press, which, however, has so far only found favor in the treatment of tolerably high grade material. However, it is possible that important improvements may yet be made in this rather forbidding direction.

In mills where there were no special facilities for slimes treatment the practice has been in some cases to allow as much dust as possible to go out of the windows, or into the lungs of the workmen. In some cases this dust carried exceptionally high values, but in going to waste it probably gave the company a better net profit than if allowed to go into the already slimy vat charge.

It is hardly necessary to mention the fact that a much greater amount of dust may be allowed in roasted material without harm. At a low red heat the water of hydration is driven from the aluminous and clayey constituents of ore. During the after treatment with aqueous solutions the water will not again combine to make a plastic or tenacious material, the dehydrated silicates having the physical character of sand, even when crushed as fine as 200 mesh.

However, we can not afford to roast simply to secure better leaching, and this brings us face to face with the problem of treating the slimes separately. As to the precise method, it is probable that we are on the right track in "leaving them alone," as much as possible, just run them into a big tank, allow the solids to settle and the gold solution to decant, then various washes, *ad infinitum, ad nauseam.* That is the most prevalent system, the one used here and today regarded the world over as good practice.

But the treatment of slimes presents an inviting field for further research, and in this the Black Hills men have not been idle. Mr. D. C. Boley has been working with his characteristic energy on the subject for the past year. He has completed and taken out a patent for a machine operated on the principle of forced percolation by air pressure while the material is thinly spread on a filter. It has many strong points from a theoretical view, but Mr. Boley, realizing that such things sometimes develop weak points in practice, has leased a small mill for the purpose of trying the invention on a working scale. Whether or not his machine is found to be an improvement over all other methods, his careful and painstaking efforts will materially add to the stock of human knowledge on an interesting subject.

The Portland mill has developed an arrangement in the bottom of its slimes tanks which introduces the wash somewhat on the principle of a revolving lawn sprinkler and is said to effect an extremely thorough agitation of the charge. It may assume the place in popular favor so long held by the more common centrifugal pump, and if so, will obviate the pumping of the charge to a different tank at each wash, which by most mill men is regarded necessary in order to make sure of thorough mixing.

It has always seemed to the writer of this article that the decantation process being the most natural and in itself requiring no machinery whatever—just let the mud stand and settle—is most likely to prove economical in practice. But as now carried on intermittently by separate tank charges, there is valuable time lost in charging, agitating and discharging, besides the labor involved, and the needless storage and rehandling of the bulky washes required. Now if we could make this process continuous, so that the entire area of the settling tanks will be continuously employed in settling the suspended material, it would leave little to be desired and certainly nothing more to be attained in decantation. It would materially cut down the floor space required, save labor and require practically no storage capacity for washes. In pursuance of this train of reasoning an ordinary pointed box settler was last fall put in at the mill of the Highland Chief Mining Company, then just put in operation by the writer as a wet crushing cyanide. The box was designed for the continuous separation of the slimes from the battery solution· It was a small affair, having only about 100 square feet of settling surface yet it handled about eight tons of dry solids per day, while the regular settling tanks of the same surface area and having a capacity of thirty metric tons liquid, required to stand undisturbed twenty-four to thirty hours to settle five to six tons of solids. An interesting feature of the pointed box was its capacity to decant an enormous quantity of liquid without retarding the subsidence of the solids. In fact its decanting capacity could not be ascertained as it was not practicable to deliver to it more than 200 tons of battery solution per day. The regular slimes tanks, having three times the cubic capacity could not turn off more than twenty tons of wash in the same length of time.

This of course suggested that a large quantity of wash can be used in a continuous apparatus, while by the intermittent process a large wash means a very big tank for a small quantity of slimes. In the operation of this box a number of important facts were observed and checked, some of them opposed to conceptions which often control the design of settling tanks for mill work, and this has encouraged the writer to undertake the design of a continuous apparatus for washing and extracting the values from slimes. It has been made the basis of patent applications, numbered 108.722, and 111,049, series of 1900, embracing some fifteen claims and containing quite a number of long words required by the rules of the Patent Office in such case made and provided. A small fraction of one percent. of all the patents issued in this country find their way into practice, and it is therefore possible, barely possible, that this apparatus may again be heard from.

This interesting subject can not be discussed farther within the usual limits of a paper of this kind, and I must therefore close, thanking you, gentlemen, for your kind attention.

The paper was very well received, and Mr. Randall congratulated by all who heard him on the excellence of his paper, and voted the thanks of the association.

The meeting then adjourned to the ante-room, where a nice lunch awaited the discussion of the members and their visitors.

PYRITE ORES AND THEIR SMELTING.

By Dr. Franklin R. Carpenter.

[Paper prepared for American Mining Congress, Deadwood and Lead, South Dakota, September 7 to 12, 1903.]

This is a process of smelting applicable to any raw ores not carrying lead, but more especially to sulphide ores carrying copper.

From time immemorable man has roasted off the sulphur in pyrite ores and burned his iron to oxide in the open air, thus wasting what pyritic smelters consider good fuel. If it is admitted that a heat unit derived from the oxidation of iron or sulphur will do as much work as one derived from the oxidation of coke, the folly of this proceeding becomes apparent, provided this heat can be utilized. If it can, one might just as well waste his coke in a similar manner.

American engineers derived from Europe two raw smelting processes, which, unfortunately, are often confused. One was the Kongsberg process of pyritic smelting, where raw pyrite was added to the charge simply to produce a carrier, or matte, for the precious metals. This was all I had in view when I advocated pyritic smelting for the siliceous ores of South Dakota. By its means the small amounts of gold and silver in many tons of rock were concentrated into a few tons of matte. This process, broadly, is very ancient; so ancient that we know not when it was first employed. It will be observed later that it is the very opposite of the other class of pyritic smelting, in that the ores treated are siliceous, and pyrite is added for a carrier only.

The other sort of pyritic smelting is the out-growth of principles discovered by Sir Henry Bessemer in steel making, who found that cast iron might be purified by the oxidation, or burning of its own contained impurities. The principles of Bessemer, much modified, are now everywhere applied to the refining of copper matte, where again the oxidation of the iron and sulphur furnish the heat to burn the slag off impurities, giving us a very pure blister copper at one direct cheap operation, and without additional fuel. This is the beautiful operation of Mahnes, first employed in America by our Butte friends.

After the establishment of Bessemer's process in England, Hollway sought to smelt the Rio Tinto copper sulphide ores by means of the heat generated in the oxidation of their sulphur and iron. A short calculation will show that his conclusions were well founded. Without going into the investigation very fully, we may admit that one pound of iron pyrite burned in the furnace is equal to 2,026 B.T. U., and that this, roughly, is equal to forty per cent of the value of a pound of carbon burned to CO; but as our furnaces probably burn perhaps a third of the carbon to CO_2, we may conclude that this value is too high, hence figure it as equal to only twenty-six per cent, or one-fourth the value of one pound of coke, which is certainly a safe deduction.

Those who are interested in the subject are referred to the forthcoming volume of the Mineral Industry, where Mr. E. C. Reybold, Jr., a young man employed at our Golden Works, and formerly with me at Deadwood has fully investigated the subject.

For every four pounds of pyrite, therefore, burned in the open air, we have lost the equivalent of one pound of good coke. Stated in another way, four pounds of pyrite will do as much smelting as one pound of coke, and in so doing, it is smelted and fluxed itself.

Our blast furnaces, in ordinary matte smelting, are running with sixteen per cent coke, but a charge containing sixty-four per cent of raw pyrite should smelt itself; and if this is assisted with a hot air stove, which can be fired with a cheap low-grade fuel, even this percentage of pyrite may be much reduced. The fullest application of these principles has been made by Mr. Robert Steicht at Mount Lyell in Tasmania, where the first smelting is done absolutely without carbonaceous fuel of any sort. Let us now consider for a moment what they do. Their ores are pyritic, and of two classes. The Mount Lyell pyrite is so mined as to maintain a general average as follows:

Fe, 40.30 per cent.
SiO_2, 4.42 per cent.
$BaSO_4$, 1.48 per cent.
Cu, 2.36 per cent.
Al_2O_3, 2.04 per cent.
S, 46.01 per cent.
Ag. 2 ounces per ton.
Au, 0.0725 ounces per ton.

The second class is a siliceous bornite ore purchased from other mines, and quartz is employed as a flux. This is the direct opposite of the case first considered, calling for additions of silica in the place of additions of pyrite.

The Mount Lyell Company operates eleven blast furnaces which are arranged in two smelting plants. Those employed in the first smelting are five in number, and are 42x210 inches at the tuyeres. The height of the ore column above the tuyeres is maintained at nine feet and six inches. The other plant consists of six furnaces, five of which are 40x168 inches at the tuyeres. The tuyeres are all three inches in diameter, and the larger furnaces have thirty-two each, the smaller ones twenty-four each. In the first set of furnaces all the ore delivered at the plant is smelted without roasting and without fuel, to a first matte carrying fifteen per cent copper. Formerly a hot blast, 528 degrees, and three per cent coke were used. But for a year past the coke has been abandoned and the blast only warmed. No difference was noticed in this change save a greatly increased capacity—three furnaces now doing the work of four under the old method.

The matte from this first smelting is re-smelted in the second set of furnaces to a forty-five to fifty per cent copper matte, which goes directly to the converters.

The process is, therefore, divided into three stages, all of which are oxidizing, and which may be said to be almost continuous Bessemerizing from beginning to end. Disregarding the time for cooling and transportation from one department to another, the time consumed from ore to copper is only six hours, and this is accomplished almost without extraneous fuel.

In the first smelting no limestone or coke is used, and but a slightly warmed blast. In the second smelting a small percentage of coke and limestone is used, and a cold blast. The third stage is simple Bessemerizing or converter work.

These results having been attained at Mount Lyell by the application of principles long advocated by pyritic smelters, there is no longer any reason, in my opinion, why the same or similar results cannot be had at Sudbury, Ontario, Ducktown, Tenn., Keswick, California, and in Arizona and New Mexico—in fact, at any place where the ores carry sufficient pyrite, or pyrite can be had from outside sources.

It will be observed that the smelting proper at Mount Lyell is accomplished in two steps. A low-grade matte is made in the first smelting, which is enriched by a second smelting to a grade high enough for the converter. This, in a differently constructed charge, may not be necessary, depending upon the per cent. of copper, degree of concentration and the proportion of iron to silica. A charge can be made of Montana ores which will not require the second, or concentration smelting; but the second smelting, being relatively small compared with the first, is never a serious matter and adds but little to the cost.

I have now sketched the two outside cases of pyritic smelting, both of which are eminently successful in their respective fields. There are many cases, however, which lie between these extremes, as at Butte, Montana, in Gilpin County, Colorado, British Columbia and elsewhere, where the sulphide ores carry a large percentage of silica and are treated by water concentration before smelting. This pre-supposes concentration mills of enormous capacity and roasting furnaces for the concentrates so obtained, both of which cost great sums of money, and which are at best very wasteful. By the further application of the principles already developed and the utilization of the cheap fuel now wasted, it is barely possible that the process might be modified.

Modern copper smelting methods have received their greatest development at Butte, and I will let no one go beyond me in admiration of the great work done there and sincere respect for those who have accomplished it. Their mills are models of mechanical ingenuity never surpassed, and their reverberatory furnace work is not equalled. I do not lose sight of the fact that these last furnaces, which a few years ago, when first introduced from Swansea, had a hearth capacity of but 9x14 feet, and a smelting capacity often as low as ten tons per day, requiring to be clayed up every twenty-four hours, have now been developed into furnaces having hearths 20x50 feet, and smelting more than 100 tons each in twenty-four hours, and which require claying not more than once in twelve days; also that they save more than fifty per cent of the fuel used in the old furnaces.

It is, therefore, with the greatest diffidence that I suggest that any change is possible in the methods of a camp which is today without a peer in the world for the excellency of its work, but let us not forget history.

"The old order changeth, giving place to new."

Seemingly small things in metallurgy have often accomplished the greatest results.

A few years ago there was still running in Savoy, a small iron blast furnace blown by a trompe, or box in which falling water compressed air by entangling it in its fall—a blower which we may readily imagine neither heated the air nor dried it, yet the addition of this blowing machine, crude as it was, made the instrument which put out of blast all the Catalan direct furnaces in every part of the world. Without the trompe the blast furnace for iron would probably not have existed, and without pig iron all that is known to us now as the "age of steel" could not have existed.

Already three-fourths of the beautiful ancient Welsh copper process, with its roastings and re-smeltings, to which the reverberatory furnace belonged, has gone by never to return—the one operation of converting having replaced them all.

I believe that when Hollway undertook to smelt the Rio Tinto copper ores without fuel other than what they themselves contained, he laid down a principle which will ultimately make every copper roasting heap and roasting furnace as useless as the Catalan forge, and the time is near at hand when one would no sooner waste his good iron sulphide fuel than he would his good coke. Already more ore is smelted raw at Butte than formerly. The first-class copper ore and the coarse concentrates go into the blast furnace raw—a tribute, as far as it goes, to pyritic smelting. But if the prinicples here mentioned are correct, the large concentrating mills and roasting furnaces will gradually be replaced by a process that is all one of fire, and that fire largely derived from the oxidation of the now wasted pyrites.

I have done what I could to secure the actual composition and cost of treating an average ton of Butte ore as it is broken at the mines, that I might make a comparison between the all-fire raw method here advocated, and the combination water concentration smelting method now employed.

The following may not be absolutely correct, but it will do for comparison. The ores of this district, according to a recent paper, are mined in two classes. The first are said to average from ten to fifteen per cent copper, and to constitute ten per cent of the ores raised. The second class comprises the remaining ninety per cent, and yields from three to six per cent copper. If all were broken down together, we may take five per cent as the average, and thirty to forty per cent silica with the alumni, alkalies, sulphur and iron to balance.

I have arrived at the present cost per ton of ore from the testimony of Mr. Frank Klepetko, in March, 1898.

Dressing (or water concentration), per ton of ore, $0.82. Roasting concentrates derived from a ton of ore, $0.38. Smelting calcines, $1.20. Total **per** ton of original ore, $2.40.

In the water concentration he stated the loss to be eighteen per cent. In the roasting, 2.6 per cent. In the smelting, 4.2 per cent. Total, 24.8 per cent.

With copper at fourteen cents per pound, this is worth $3.47, making the total cost, including losses, $5.87 per ton of original ore.

·If this ore were smelted direct as it comes from the mine, without concentrating or roasting, by the addition of limestone and coke and the application of hot blast, it would cost fully as much per ton of ore, perhaps more; but I am sure that $3.00 per ton will cover it. This is more per ton, but I estimate a greater saving. According to the above statement there was lost, in the concentrating, roasting and smelting 24.8 per cent of the original contents of the ore by the time the copper was raised to a grade sufficiently high for the converters—against which I figure but nine per cent in direct smelting, leaving a difference of $3.47 minus $1.26, equal to $2.21 gain per ton. As this gain is wholly in the copper, it adds a proportional length of life to the mines. If, however, the one smelting cannot be done for the cost of concentrating, roasting and smelting, this gain would be reduced by the difference. If the first smelting cost $3.00, which I am sure is ample, we should still have a gain of $1.60 in favor of raw smelting, always supposing these figures to be correct.

In our prejudice for the established methods, it will be well to remember a story told by a traveler from the Sahara Desert. He came across a party of Arabs making iron—doubtless after a manner dating from the days of Abraham. The furnace consisted of a hole in the ground, around which were three blowing engines, each consisting of an Arab with a long tube, one end of which was in his mouth and the other in the furnace. After blowing, from six to eight pounds of iron per shift was obtained.

Our traveler was much impressed, but inquired of the boss metallurgist, "Why do you use this method of making iron?" He received a look of withering scorn and the reply, "What other method can there be? Neither our fathers nor ourselves ever heard of any other."

Because our fathers and ourselves have always burned our iron and sulphur outside of the furnace is no good reason for continuing it.

GOLDEN REWARD SMELTER, DEADWOOD, S. D.

MATTE SMELTING.

By Paul Danckwardt, Superintendent Golden Reward Smelter.

[Read before Black Hills Mining Men's Association, July 24th, 1902.]

Gentlemen:

The process of matte smelting, which I have chosen for the subject of this evening's paper, is rather an old affair for the Black Hills people; it is practically one of the first processes introduced here to solve the problem of extracting the values from our ores. But however old the process of matte smelting, often perversely called the "pyritic process," may be, it has doubtless achieved vast results, and on account of this and the fact that this process generally everywhere else is in a state of rapid improvement and expansion, it may be worth while, to state here what we have done in all these years, what we are doing and what can be done for the future of this process as far as the Black Hills conditions are concerned.

As this process of ore reduction is only by one company successfully applied in the Black Hills, there is no competition in this field of metallurgy right here. And consequently no basis for exhaustive discussions of the pro and con, as we used to hear them with regard to the cyanide process. The cyanide process forms indeed at present the topic of the day, but both processes have their special field of application. Taking all the different classes of ore into consideration, we may shortly say, that both processes together will obtain the best result.

When matte smelting was first introduced in these Hills, there seemed to exist a confusion of ideas as to the name, under which the "new" process was to be launched. Even patents were taken out, but at last the truth dawned, and at present we hear nobody claim to have invented this good old process, called matte smelting, because it is too old to have parents living in our generation. But this statement shall not say that we are doing today exactly the same thing that has been done a hundred years ago or that no exertion has been made to change or improve the old way of working this process. On the contrary, there have been expended thousands of dollars and lots of brain power to make out of the good old process with limited applicability one that is now fit for competition on a widely enlarged field. These changes are not fully recorded in our text books and the lay man or merely book-learned individual hardly conceives the extent of the improvements made since its introduction into this country. Today even, we cannot state yet that the time of transition for this process has past already.

If we begin now with the first question: What have we done since the old D. & D. smelter started? I may say, that we have tried everything. One experiment followed another, and credit to the capitalists and the managers for their persistency in their work and money consumed. Soon we would make an iron matte, soon a lead matte, soon we smelted lead, soon iron, commonly called "sow," and someone considered that the only possibility of solving the problem. Slags were changed and the furnaces tested for what they could and could not do, until they were frozen solid from top to bottom. Muscular smelting played an important factor during the infancy of the works. At last we settled down on a slag containing about forty per cent silica, thirteen per cent. ferrous oxide, and the remainder alumina, lime and magnesia, the lime prevailing. We could now at least adjust our coke to a certain amount of material to be smelted, and often we stuck too tenaciously to the sogained rule, sometimes to the disadvantage of the company financially. For there were other points to be settled, which had just as much bearing on the running of the furnaces as amount or quality of the coke and slag forming material alone. When during the first years of the operation of the works an iron matte was made, the furnaces sometimes ceased providing this desirable compound suddenly, and consequently the slags would run up in gold and silver values, there being no carrier for them. The men, even the head men, used to say, "the matte is hanging up somewheres," because they noticed sometimes afterwards, that the furnaces suddenly changed and would start to make such an amount of matte, that neither furnace nor forehearth bottom would stand the corroding action of the gushing iron compound. If this was the end of it, there remained nothing to be done but jump and the let elements cool down again. But more often it happened that the tragedy closed in the form of a gradual freeze up. The matte actually did not hang up to any great extent at the beginning of the freeze up, and might have been forced to reappear, as the iron compounds of the charge, principally pyrites, were simply burning off, there being no proper proportion between air blast and coke. When thus the formation of the matte ceased, the gold and silver of course could not do anything but run off with the slag. This was a great drawback, but happily this condition of affairs could not exist very long, because the furnace, as if it had a soul, that cannot see the values spoiled in such a way, would either, if conditions were favorable, correct itself, or cease to work altogether. This latter was to be the end of it, if the matte did not start to run again, as there was then nothing running down to keep the furnace bottom hot enough for regular work.

Both these often occurring freeze ups and the temporary appearance of a high slag, had to be avoided in the future. Therefore experiments were made with a copper matte. The good effect of this was evident, but I for my part would have looked for other means of obtaining the same result, as the use of copper ore is at times a great expense, which is not always outweighed by the gain in saving of valuable metals. Since I have had charge of the plant I have therefore cut the copper down to a few per cent and obtain the much desired regularity of flow of matte by giving at intervals, in accordance with the general condition of the furnace, a charge high in iron and of great heating power.

Up to this time, when the change from iron to copper matte was arranged, the furnaces were run at a low air pressure, the output of the mill being consequently small. The following introduction of an increase of pressure raised the capacity considerable.

The same effect had the installation of larger furnaces. But in spite of all these improvements the furnaces often happened to make only short runs. Often a freeze up would occur right after the blow in, and the men at the head could not always give an explanation for such accidents. It should have been evident, that a furnace running at such a high silica and low iron slag, has to be treated in an altogether different way than a copper or lead furnace when blown in. At this time, when the bottom is cold, the jackets not yet or only slightly crusted with slag, and the charge loose, there is great danger from losing heat. The furnace ought to be watched very closely, as otherwise the coke will burn off, before the charge is melted, and the partially melted charge enwrapping part of the coke will fill the bottom up rapidly. This is the time when the furnace man can show his capability. It is sometimes hard to find the fault before it is too late, the only indications being the condition of the tuyeres and the appearance of the top of the furnace and the temperature of the jackets. There existed formerly the idea, that it was dangerous to try the tuyeres right after blowing in, and even later on, when the bottom was up to its full heat, the punching was not resorted to, except a tuyere would fill up with molten slag. This was altogether wrong. At present the tuyeres of all the furnaces are tested every morning in the presence of the superintendent, and, if deemed necessary, even out of the time. Only a careful examination will make a correct diagnosis and the application of the proper remedy possible. The condition of the top will show best, whether there is too much or too little coke, when taken into consideration together with the condition of the tuyeres. Any bad tuyeres should be turned off right away; if the number is too great, only part of them, preferably those adjoining some tuyeres having good fire, in order to give the fire a chance to get down again to the tuyere level. The jackets should in the beginning always be rather warm.

These points never have been cared for enough up to a couple of years ago, probably for want of the right conception of their importance. Some queer phenomena might have found an easy explanation, if the above rules had been applied. At present we can say, that a freeze up is a matter of the past; what finishes up a furnace or necessitates a blow out, is either a hole burned in a jacket, which occurs probably once in years, or a leak sprung in a coil. Someone of you gentlemen, who has seen or had a more intimate acquaintance with blast furnaces somewheres else, may ask for an explanation of a coil, because he did not see any of the kind. Well, first this coil is almost invisible, being imbedded in the heavy brickwork of the upper furnace wall, then there may be another reason, why he did not see it, that is, because it actually did not exist. A coil is really a thing of the past, a constructive rudiment of a century ago. Already before the old D. & D. smelter was put up, other plants had the coils replaced by water jackets running clear to the top. Nevertheless, when the fire destroyed the D. & D. plant, and new furnaces were bought, it was decided to stick to the coils. Some experiments had been made with wrongly constructed top jackets, which proved a failure, mal-conducted experiment in combination with an idea of the management, that the water, running even through small coils would have a tendency to freeze up a furnace, caused such a wrong conclusion. Yes, it has happened, that orders were given to turn the water off from a new coil, because it seemed to threaten a freeze up. It was a saying, that the coils caused a loss of heat, and therefore a big coil, which had been tried because the small ones burned out easily, and thus finished the life of the furnace, was given up again. When I took charge of the mill I took the matter up again, and today we are running even a bigger coil than ever was tried before with perfect success. At least by means of this coil we can now run a furnace over six months. By that time there has accumulated so much mud in the lower part of the jacket, that a blow out is imperative, or you will run the risk of an explosion. Lately I have put in a contrivance, which will, I hope overcome also this obstacle as it will afford a means to remove any scale or mud from the jackets without blowing the furnace out. If this works all right there would be nothing to limit the life of a furnace but a terrible disaster to the driving machinery. Small break downs of engine or pumps do not finish the run of furnaces. We have had last winter shut downs of forty hours on account of shortness of supplies, at the time of the snow blockade, followed by other shut downs a day later, which ended none of the furnaces concerned. Formerly a shut down for some greater length of time was considered almost a sure disaster to the furnaces.

I have spoken so far only of the one department of the Golden Reward smelter, which is at present running. But up to a little over a year ago, there used to be in operation a number of reverberatory furnaces. These were put in when the Homestake Company shipped their concentrates to us. At that time there was formed such a vast amount of flue dust from the blast furnaces that it was hardly possible to run it over again through the same furnaces. Before the installation of the reverberatories, however, it was tried to lower the amount of flue dust formed from these fines by mixing them with lime or molasses but without success. The reverberatories proved to be the best way of treating them, though they caused a big extra expense, but taking the good effect of the concentrates on the furnace run into account, it paid to do it. At present we are not getting any more concentrates, and having only little flue dust, which we can handle easily by running it over again in the blast furnaces, the reverberatories are shutdown.

These reverberatory furnaces were, however, used not only for the treatment of the flue dust, but also the sows formed in the blast furnaces and the forehearths, the amount of which was at times very great, were handled in them. Both the formation of such sows to such an extent and the separate working of them was in my opinion no necessity and certainly not a paying institution. I found out that they form, if judiciously given back to the blast furnaces an excellent material for keeping the same in a first class shape. But just as good it is to reduce the formation of them to as little as possible, as they carry a high value of gold, which has to be rehandled. I obtain this result by widening the furnaces at the tuyere level. Our furnaces, when I took hold of the plant, were from thirty-four to thirty-six inches wide, at present none is narrower than forty inches. This prevents the reduction of iron to the form of sow sufficiently.

Gentlemen, having thus arrived almost at the end of our review of the past and present of the works, we have to enter now a question, which I do not consider as settled yet. By relinquishing the Homestake concentrates we had to resort to a kind of pyrite, the only one available in large quantities here in the Black Hills, but a very poor substitute on account of its graphitic character. You all know, that graphite is a modification of carbon, that requires an immense temperature to burn off. In the beginning we had lots of trouble with them; I tried to smelt them preparatory in the reverberatories, but found that they could not be melted at the temperature obtainable in such a furnace. When put directly into the blast furnace in the shape they were received in from the mine, they caused the charges to become extremely flamy on top, so that the feeders could hardly do their work properly. If then even the coke happened to be rather dense or too much broken up, the evil was much more. I noticed also, that the coarseness of the charge generally had an unfavorable influence on this phenomena, and as something had to be done, I decided to make some experiments with crushing the pyrites, which scheme led at last to some degree of success. They continue, however, to be a hard proposition, not only on account of the fires rising to the top once in a while, but the whole run of the furnace, if we compare it with that made with concentrates, is impeded.

Gentlemen, we will now consider the improvements that can be made on the process and apparatus of the present plant proper. As stated above, the substitution of a more suitable kind of pyrite for the graphite bearing material is much desirable, and the supply, if possible, should be increased to such an extent, that no or nearly no lime would be required as a flux and the coke charge could be cut down to about one-half of the present amount. Such a change would avoid the addition of copper ore altogether and cause a greater saving in values. As to the apparatus, though little things have been changed gradually with the run of lime, as I have stated, no radical changes have been made.

If all these changes are made, the reduction in treatment expense will be about fifty per cent of the present. I hope that the next year will realize some of them already, I am well aware what a great responsibility must be incurred by the management, in entering all these questions for adjustment, but enterprise assisted by good judgment is the only spring that turns the wheel of chances for ultimate success, and can give the mill, that has successfully competed for all these years with the other ore reduction plants, and outlived some of them, a good footing in the struggle for its further existence. With the ample supply of ore, which the Golden Reward Company keeps on hand, to me the future of the smelting works looks brighter than ever before, and being at the same time in charge of the company's cyanide mill, I am probably in a position to weigh off results and I will close my remarks by saying, that both processes promise to hold their own fields here in the Black Hills.

MINING IN THE BALD MOUNTAIN AND RUBY BASIN DISTRICTS OF THE BLACK HILLS.

By John Blatchford.

Gentlemen:

In describing a portion of this formation I shall not touch on the geological part of it because that has been gone into so extensively by such men as Newton, Devereaux, Headden, Blake, Jenny, Carpenter, Hoffman, Farrish, Dr. McGillicuddy, Rickard, Smith, Fulton, O'Harra, and a number of other noted men who have written some very good papers showing the geological features of this part of the country. I merely intend to say a few words on the occurrences of the ore bodies as we find them in this formation.

These ores were first discovered in 1877, but there was very little done on them until 1890 and 1891, because, up to this time, all of the ore had to be hauled by teams and shipped out of the country to be treated. In the latter part of the summer of 1891, the Burlington and the Elkhorn Railroad Companies placed a number of spurs into the different mines; after this the work really began in earnest.

At this time it was not known how extensive these ore bodies would prove to be, but after continuous work for over twelve years, now, we find that they are almost unlimited. Ores that we could not look at years ago. on account of their low grade, can be handled today, with our new reducing or cyanide process, at a profit.

Since it has been discovered what these ores can be treated for with this process, we find that we have to work over a considerable area of that which has already been worked. There is no doubt but that this will be a great advantage to us, in the future, because we will be able to take out our low grade ore, as well as the better grade, as we advance in our work.

The ore bodies or chutes are numerous. The largest bodies so far discovered, of the better grade ore, east of Bald Mountain and Terry's Peak, lie on the quartzites, and these lie on the Archaen schists and slates; this is what is known as the vertical formation. Some places in our mines the flat ore body is known to lay and to be intermixed with a vertical ore body, which comes from below, not showing any division by quartzites. It is one of those occurrences which causes me to believe that there are a number of those vertical ore bodies, or quartz ledges, that are covered up, by this sedimentary formation, for instance, quite a portion of the Homestake ore bodies have been more or less covered by this flat formation, but in other places the flat portion being more or less eroded, left the vertical portion to be more easily prospected than it is in this district.

The eastern boundary of the flat formation begins at the original Golden Reward and Buxton, and almost at the base of Bald Mountain on the north and to the west of the Sugar Loaf Mountain on the south. It starts with a thin layer of quartzite, lying on the schist, covered with sandstone and shales; it gradually thickens towards the west, not so much because the hill rises but because the quartzite and schist drop. It drops at various distances, at a time, until it gets several hundred feet below the surface; making a number of layers of different material above it and on and between some of these layers is where we find what is called top contacts.

As we get nearer Terry's Peak the flat formation thickens more by the rise of the surface than by the fall of the quartzite, and west of the Peak it seems to keep this thickness for a number of miles. Towards this rise or thickening of the formation is where the top layers of ore become more numerous. How many layers or so-called contacts there are has not yet been determined. There is something new continually cropping out.

In these upper layers we usually find a vertical or crack filled with ore extending downwards for hundreds of feet, with a number of lense-like shaped bodies of ore, branching out at different intervals, some places connecting with bodies from nearby verticals.

At present most of the workings west of the Peak are on the upper contacts. In the Ragged Top district the ore bodies are up in the lime and they are proving to be very extensive and profitable. Around Portland they are all in shales, scarcely any work in that neighborhood being done on the quartzite as yet. There is no doubt in my mind when they commence to look for the lower ore bodies west of the Peak but that they will find them large and valuable on the quartzite just the same as they occur east of the Peak.

The gulches on the surface on the east side of the mountain all trend toward the east and on the west side towards the west, but underground we find this different; from Bald Mountain south it appears that the original channels all flowed to the south and from the north of Bald Mountain to the north. The water courses and the dip of the quartzite show this to be the case. Present conditions are exactly the opposite of the original conditions. The original dykes all have a north and south course, while a few of the later dykes near the base of Bald Mountain have an east and west course and the ore bodies, or nine-tenths of them, have a north and south course.

These ore bodies vary in width and thickness; we find some of them over four hundred feet in width and various thicknesses, from six to twenty feet, and of various values, ranging from five to fifty dollars. The general average of what we call smelter ores are about twenty dollars per ton and a general average of cyanide ores in the neighborhood of from eight to ten dollars per ton.

To describe the conditions of the quartzite we may compare them with the waves of the ocean. Some places we might imagine there was not much wind, making the quartzite smooth and then a big wind lifts the wave up from two to three hundred feet, the quartzite raises the same, some places we have one hundred feet, from that to two hundred feet or more across the top of it going down some places almost at a vertical, or some places with a gradual slope, others with steps.

We find these ore bodies at the base, on the steps and slopes, most times on the top of these large up-lifts, but very seldom find any ore bodies in the channel proper. And it appears that the most of the level places in the quartzites seem to be capped with large sheets of porphyry, but at every fault and in close proximity with a fault. The capping is most all composed of shales and sandrocks. No doubt this has a good deal to do with the occurrences of the ore along the breaks, those being in themselves an altered condition of these same shales and sandrocks.

There is no question but that this flat ore formation follows the lime stone ridge from between fifty and sixty miles on the south and about twenty-five miles on the west, and to Spearfish on the north.

This does not include all of the flat formation of the Black Hills. The Galena district has a very extensive area of this formation. The present developments there are very encouraging, although there has not been enough done to determine how large the ore bodies are, but they are numerous, and prospects obtained from most of them are good. There is still a very large area in those districts undeveloped.

There is room for a good many mines such as ours, which is the Golden Reward Mining Company's property, consisting of over fifty miles of underground workings, about two-thirds of this mileage being on ore channels, while the other third is cross cutting barren rock to find ore channels. After following some of these ore bodies close on to three miles we find them still continuous.

THE GEOLOGY AND MINERALOGY OF THE BLACK HILLS REGION.

By Cleophas C. O'Harra, South Dakota School of Mines.

The Black Hills region is in many respects a typical geological unit. It lies within the forks of the Cheyenne River on the South Dakota-Wyoming boundary line, a much larger portion of the area being within the state of South Dakota. Separated from the Rocky Mountains to the west and southwest by a distance of less than 150 miles the region possesses many of the lithologic and physiographic features of that great mountain system.

Structurally the region is an elliptical, outwardly-dipping uplift, the more distinct features of which cover an area about 100 miles long and fifty miles wide, the longer axis approximately coinciding with the meridian except in the northern portion where the general direction is to the northwest. By reason of its isolated position, its simple structural features and the many excellent natural and artificial rock exposures, the history of the region may be interpreted with a considerable degree of ease.

The general system of drainage is distinctly radial. The two enclosing arms of the Cheyenne river wholly separate the Hills from other drainage systems and receive the many smaller streams from the more elevated mountainous portions, a high western limestone plateau being the main divide. In certain places the rapid erosion of softer beds has modified this general radial arrangement, a prominent example being in the formation of the well-known Red Valley, which forms a nearly continuous encircling depression separating the higher central portions of the uplift from the distinct but less elevated cretaceous hogback ridges of the foot hills.

Many of the streams continue actively cutting their beds. Each lithologic unit with its particular and sometimes striking color yields distinctive topographic forms dependent upon relative capacity for resisting erosion, the result being that in many places features of rare interest are produced. The Harney Peak area of the Southern Hills with its bold pinnacles and walls of coarse bare granite rising from their forest-clad base of metamorphosed sedimentary rocks presents a beautiful panorama, while the steep-walled canyons of Spearfish Creek, and of Elk Creek in the Northern Hills, are among the most picturesque that America can show. Again to the northwest along the Belle Fourche valley where the horizontal sandstones and shales have been intricately carved by the various streams, and where the brilliant and varied colors of the several formations harmoniously blend with a wealth of forest and pasture overlooked here and there by the stately, somber forms of porphyry buttes, there is presented a view well worth many a hardship to see.

The prominent topographic features are a high central basin of granite and metamorphic rocks of Algonkian age, surrounded in a concentric manner by a rugged infacing escarpment of massive, white, carboniferous limestone, a wide depression in the red Triassic shales and a high rim of Cretaceous hogback ridges or foot hills. Beyond these are the later Cretaceous shale formations which give rise to the nearly level plains. Farther away on almost every side, interrupting the otherwise monotonous approach to the Hills, there are abrupt tables and buttes of Tertiary clays, large portions of which have been carved into forms that bewilder the imagination of the most fanciful observer.

In the Northern Hills Tertiary intrusive rocks have greatly modified the general topography, and in not a few instances have formed prominent landmarks. Terry Peak, situated near the center of activity of intrusions, is the highest point. It reaches an altitude of 7,069 feet. Some distance to the west of this is the Bear Lodge range which culminates in Warren Peaks, marking a subordinate but important center of Tertiary disturbance. Several isolated igneous peaks differing little in petrographic and structural nature from the prominent peaks of the more intricately disturbed districts, already mentioned, stand as tall sentinels among the lower peripheral ridges, chief of these are Bear Butte, Crow Peak, Black Buttes, Inyan Kara, the Missouri Buttes, and the justly-famed Devil's Tower. The highest point within the entire region, as it is also the highest point in the United States east of the Rocky Mountains, is Harney Peak. This is the culminating peak of the Harney granite range in the Southern Hills. It reaches a height of 7,216 feet. The surrounding limestone escarpment rises high above much of the inner portion of the Hills, and considerable areas of the plateau along the western side in the vicinity of Crooks Tower closely approach the height of Harney Peak. The mean altitude of the plains surrounding the Hills is little more than 3,000 feet. The average elevation within the hogback ridges is approximately 5,000 feet.

The rocks of the Black Hills show a wide range in age and character. Within the crystalline nucleus are pre-Cambrian granites, amphibolites, schists, slates, phyllites. and quartzites. Beyond this nucleus are limestones, sandstones, shales and conglomerates representing a nearly complete sequence from Cambrian to Laramie. Their combined thickness is approximately 10,000 feet. Extensive overlaps of Tertiary rocks are also present while Pleistocene deposits of various kinds occur widely distributed over the surface of the region. Silurian limestone is found in a few localities, but is of little importance. The presence of Devonian rocks seems as yet not conclusively proven. In the Northern Hills there are porphyritic rocks in great profusion. Phonolites, Grorudites, Audesites, Dacites, Diorites and Lamprophyres are found and their recent careful study has aroused much interest among petrographers. To the prospector and miner they are of interest in that their intrusion has greatly influenced mineralization, and the nature and distribution of the igneous masses have to no little extent been a determining factor in the occurrence of ore bodies. Fossiliferous beds are common among the foot hills, while only a short distance to the southeast are the world-renowned White River bad lands, with their wealth of vertebrate remains.

The sedimentary deposits were laid down subsequent to the upturning and metamorphism of the Algonkian rocks. These have had their various characters properly defined by recent study, the determined formational units receiving appropriate individual names. The oldest rocks of the region are the slates, schists and quartzites. They constitute the main central area of the Hills. Their dip approximates the vertical, while their strike corresponds fairly well in a general way with the meridian line. The quartzites are usually less easily eroded than the slates and schists, in consequence of which they not infrequently stand out with much prominence. Dark, basic, igneous bands occur in many places, their general occurrence being such as to give the impression of intercalation conformable to the original bedding of the metamorphosed sediments. These rocks have not received careful study, but they may be provisionally grouped under the name "amphibolites." They are commonly designated by the prospector as diorite or hornblende rock. Intimately associated with all of these are the granites of the Southern and Central Hills. In the northwestern part of the Hills on the South Dakota-Wyoming line, another small but important area of granite is found. A distinct feature of nearly all of this granite is its extremely coarse texture. Its feldspar, quartz and mica, and even the less important and non-essential constituents may be frequently found in isolated crystalline masses of great size. The rock is of the variety of granite known as pegmatite, and as usual with pegmatite, carries an abundance of rare and useful minerals. Following the granites, which are later than the amphibolites, but still of Algonkian age, there were no igneous intrusions until the Tertiary. Then approximately coincident with the general uplift of the Black Hills region, came the igneous bodies so abundant in the Northern Hills. These, for want of a better collective term, are commonly designated as porphyries. They are generally, although not always, of a distinctly porphyritic nature, the large crystals being quartz, or more frequently some form of feldspar, or occasionally hornblende or biotite. To mention all of the localities where these may be found would be a tiresome task. The following important mountains must suffice: Terry Peak, Bald Mountain, Elk Mountain, Ragged Top, Devil's Tower, Custer Peak, Bear Butte, Crow Peak, Inyan Kara, Sundance Mountain, and Warren Peaks. Less prominent masses occur in great profusion, and few important gulches of the region are free from good exposures where structural details may be determined with much precision. The intrusions occur in the form of dikes, stocks, sills and laccoliths, few regions in the world showing them in greater number or to better advantage. Intermediate and connecting stages of every grade are found, and erosion has planed and dissected the rock-masses so carefully that the faithful observer may easily read their meaning.

Reviewing and collecting the foregoing facts with reference to the sequence of occurrence of the many phenomena, it may be said that in Algonkian time the schists and quartzites were deposited as sediments derived from some unknown Archaean land-mass lying apparently either to the west or to the northeast of the position now occupied by the Hills. Later these sediments were penetrated by basic eruptives and subsequent to this penetration the sediments as well as the basic eruptives were ramified by quartz veins, many of which are gold bearing. Following the eruption of the basic rocks, and after most or all of the gold-bearing quartz veins were formed, extensive granite intrusions occurred. At some time during these disturbances, great metamorphism took place, the slates and the schists reaching much the condition in which we now find them. During the middle or latter part of the Algonkian period, the sea shallowed and the land rising above the sea, as an island reached a considerable height. The rocks thus brought under the influence of erosive agents, supplied much or all of the sediments which make up the Cambrian strata. After this the land became submerged, the later Paleozoic and the Mesozoic sediments indicating at first deep water, followed again by an unsteady tendency toward shallowing of the sea.

Near the beginning of Tertiary time, great disturbances took place. The region was lifted quite above the sea and deeply cut by outflowing streams. Sea conditions disappeared, leaving the land partially or wholly surrounded by a considerable body of water in the form of a lake. Approximately coincident with these changes, the porphyritic rocks of the Northern Hills were intruded and by their subsequent denudation and degradation added their portion of sediments to the surrounding lake. The lake then disappeared and upon its dry bed the modern streams have trenched their way.

The unravelling of all these facts is a matter of much interest to one desirous of knowing the processes of nature's activities, and doubly so to him who seeks for mineral wealth. The Black Hills rank among the important mineral producers of the country. Among the ores and minerals already productive, or giving promise of production, the following are of importance:

Gold,	Mica,
Copper,	Spodumene,
Iron,	Building Stone,
Manganese,	Brick Clays,
Silver and Lead,	Gypsum,
Tin,	Coal,
Tungsten,	Petroleum.
Graphite,	

Of all these, gold is preeminently the chief product. Its presence may be detected in almost every variety of rock within the region, and workable bodies of ore are found in several different formations. The following classification gives the various horizons and indicates the mode of occurrence of the ores:

A. Ores occurring within the Algonkian rocks.
  1. In quartz veins.
  2. In veins of auriferous pyrite.
  3. In igneous dikes, sheets, etcetera.
  4. In slate breccias.
  5. In fissure veins.
  6. In mineralized zones.

  B. Ores occurring within the Cambrian rocks.
      7. In the basal conglomerate—"cement ores."
      8. In the slates, sandstones and quartzites—"siliceous" ores.
  C. Ores occurring within the Carboniferous rocks.
      9. In brecciated "verticals" in limestone—"siliceous" ores.
     10. In massive limestone—"lime-siliceous" ores.
  D. Ores occurring within the Pleistocene deposits.
     11. In high level bars—"dry" placers.
     12. In present stream beds—"wet" placers.

Of these deposits, the placers, the cement ores, and the brecciated limestone verticals early yielded their most profitable returns; the pyrite veins have been extensively exploited only as a source of fluxing material for smelting operations, while the igneous dikes and sheets, slate breccias, etcetera, although occasionally of importance, have not received thorough attention. The gold-bearing quartz veins are found throughout the highly metamorphic area of the Hills, Custer, Pennington and Lawrence counties all showing localities yielding good values. The ore is generally free-milling, but there are certain important exceptions, the occurrence of which has hindered successful development.

The siliceous and the lime-siliceous ores so extensive in the Northern Hills, are wholly refractory. They occur in the form of shoots or channels in immediate connection with nearly vertical fractures, running in a direction parallel to the longer axis of the shoots. These fractures or verticals, as they are frequently called, are generally slickensided and frequently form fault planes along which more or less movement has occurred. The ore shoots vary considerably in shape, but in the main are greatly elongated bodies having a rounded or lenticular cross-section. They lie in a general north-south direction and, excepting certain irregularities produced by lateral branches, are practically parallel with each other. The structural relations are occasionally complex. Folding is observed, faulting frequently occurs, and igneous intrusions sometimes aid in concealing true stratigraphic relations. Usually, however, the conditions are of such a nature as to cause no serious hindrance to the proper development of mining property. The siliceous ores are found at various horizons within the Cambrian, chief of which is immediately above the conglomeratic quartzite. The lime-siliceous ores occur at various horizons within the Carboniferous, the chief position being near the top of the massive white or gray limestone, now technically known as the Pahasappa formation. Tertiary igneous rocks have cut and intercolated the Cambrian and Carboniferous strata to a marked degree, and it is to this action either directly or indirectly, that the deposition of the ores is due.

Of all the classes of ores mentioned, that of the impregnated zones has longest yielded large returns. The typical zone, the Homestake Belt, has furnished approximately three-fourths of the total gold output of the Hills, and continues today to afford more than one-half the annual production. The ore is chiefly low grade and occurs in extensive deposits as garnetiferous amphibole schists highly impregnated with quartz. It is largely free-milling.

Of metallic mineral products other than gold in the Hills, copper, iron, manganese and tin have received much attention, but as yet no properties worked for them have become steadily productive. Copper is found chiefly in the Algonkian rocks and nearly every portion of the Hills discloses its presence. Many of the properties are capped by a heavy gossan, carrying more or less copper and in various places where this gossan cap discloses considerable quantities of copper extensive prospecting is being carried on. As usual with such deposits, carbonates, oxides, and the native metal are found near the surface, while below sulphides occur. A zone of enrichment which, judging from other regions showing apparently similar conditions, might be confidently expected, has not yet been disclosed. Only future extensive prospecting under favorable conditions can prove conclusively the actual conditions.

Iron is widely distributed, and in the central and southern Hills in association with the slates and schists, it has received some attention. Distance from ready markets have thus far prevented its extensive exploitation.

Silver and lead are found in the Algonkian metamorphic rocks and in the Cambrian and Carboniferous sedimentaries. Lead ores associated with silver have been mined at Spokane in the Central Hills, and at Carbonate and Galena in the Northern Hills. The character of the ore bodies in the several localities varies widely. They occur in the Algonkian in veins, in the Cambrian as shoots, and in the Carboniferous as contact deposits, the latter two graduating more or less into each other. In the Central Hills the ore is closely associated with vein quartz. at Iron Hill it occupied a nearly vertical position along a porphyry dike, where the latter cuts the massive Pahasappa limestone. At Galena the ore bodies are found with the Cambrian, their manner of occurrence being much the same as the Cambrian siliceous gold ores. They, like the gold ores, are impregnations due to water which has gained access to the easily replaceable calcareous materials through numerous vertical cracks or fissures produced by the intrusion of Tertiary igneous rocks.

Tin is found in the granites and in the stream gravels of the Harney Peak and the Nigger Hill districts. The ore occurs in the form of cassiterite. Cupro-cassiterite occurs at the Etta mine near Keystone, and stannite has been identified, but these last, aside from their scientific interest, are of no value. The cassiterite is found as a constituent mineral of the granite in crystals or masses of all sizes up to occasionally several pounds weight. The mineral occurs chiefly in a feldspar-muscorite aggregate, but is sometimes found in a quartz-muscorite aggregate or in quartz or feldspar alone. The granite is generally of a distinctly pegmatitic character, and where the tin occurs is in the nature of dike material. The wide distribution of the cassiterite is readily conceded, but the actual value of the deposits is a much discussed problem, the nature and details of which have been so often touched upon that there is no need in this paper to offer opinions upon the subject. It is proper to say, however, that extensive exploratory operations are now being carried on, the results of which are awaited with much interest.

Wolframite or tungsten, as it is frequently called, is one of the most recent minerals to enter the list of Black Hills metallic products. This has been long recognized in small amounts in the granites in Pennington and Custer counties, but four years ago it was found in quantity in Lawrence county closely associated with the Cambrian siliceous gold ores. The chief occurrences are near Lead and Yellow creek, from which places several carloads have been shipped.

This extremely brief review of the metallic minerals brings us to the non-metallic products. These are of great interest and they might readily lend themselves to extended discussion. There is opportunity here, however, to give them little more than mention. Graphite occurs in the slates and schists in uncertain quantity and value. Mica in the Harney Peak granite area has long been worked and still receives prominent attention. Spodumene, also in the Harney Peak granite, especially near Keystone, is extensively worked for its lithia content. It is well to state, and of interest to remember that this mineral occurs in crystals of unprecedented size no other place in the world so far as known, showing crystals of any substance comparable in size to the Spodumene crystals of the Keystone district. Building stone is abundant. Few of the geological formations are wholly lacking in materials fairly suitable for building purposes, and several of them can supply good stone in unlimited quantity. Thus far the Dakota, the Lakota, and the Unkpapa sandstones have received most attention. The stone is readily accessible, is durable and is otherwise suitable for structural purposes. Brick clays and gypsum are abundant and easily secured. Coal is found in the lower part of the Lakota along the western and northwestern part of the Hills in Wyoming, it having been mined for some years near Newcastle and at Aladdin. Petroleum occurs in some quantity in the Benton shales in the vicinity of Newcastle, and indications of it are found elsewhere in the foot hills.

In concluding this brief summary of the geology and mineralogy of the Black Hills region, I would add testimony to that of many another before me that the Black Hills region is truly a land of wealth and beauty, a most interesting part of nature's great storehouse where men may seek with profit the material necessities of life, and where they may not find lacking those things which gratify the mental nature and which tend to lead to truer living.

GEOLOGY OF SOUTH DAKOTA.

Dr. J. E. Todd, State Geologist.

The task asked of me is to give, so far as is practicable in the time allowed, a sketch of the geology of our state, particularly of that portion outside of the Black Hills. The Hills being more complicated and not perfectly explored, I cheerfully leave to others who have more time to devote to its colaboration and presentation. Moreover, as you can readily understand, we have only time to select some of the more salient features of the vast amount of details necessarily connected with such a theme.

It will be my aim to present in order the various geological formations, give their leading characteristics, their extent, and note their more important economic relations. As few of them have to do with mining enterprises directly, I shall assume some freedom to go beyond the strict aim of the congress and shall venture to bring in a few facts not directly connected with mining.

After a discussion of the geologic map I will present illustrations of different formations by the help of the stereopticon.

GENERAL STRUCTURE OF STATE.

For the benefit of those unfamiliar with our state I make a few general statements which may seem trite to those already acquainted with it.

South Dakota presents greater range of altitude and greater variety of topography than any other state east of the Rocky Mountains. Its lowest point, Big Stone Lake, is 967 feet above the sea, and Harney Peak, its highest, 7,215 feet. It has extensive plains rivaling a floor in smoothness, rugged mountains surpassing anything in the Appalachians, buttes rising like giant pyramids above the plains, and weird bad lands, the veritable work of goblins.

South Dakota has also a greater variety of geological formations than any other state east of the Rocky Mountains, presenting a nearly complete series from the oldest to the youngest rocks.

It has two centers of ancient crystaline rocks at opposite ends of the state. Around one nearly all of the Paleozoic formations circle, and against the other most of the Mesozoic rest, while the Tertiary rocks lie between, and the Quaternary deposits are developed in wonderful profusion both of acqueous and glacial origin.

Late estimates of the thickness of sedimentary or stratified rocks in the state reach a maximum of 10,500 feet, of which 1,300 are Palezoic, 8,000 Mesozoic, and 1,200 Cenozoic. If we make the bottom of the Creataceous strata, or the crest of the "Hogback" surrounding the Black Hills, the dividing line, there will be about 8,000 feet of sedimentary rock outside of that limit, and about 2,000 inside. In this, of course, it will be understood, we neither include the schists, granites or porphyries of the Black Hills nor the granites and quartzite of the eastern end of the state, which together are commonly estimated to have a thickness two or three times as great.

The stratified rocks outside of the Hills consist mostly of soft shales, clays and sands, though extensive deposits of sandstone and limestone appear in some localities. The general softness of the strata is attested by two-inch holes being drilled 2,000 feet in depth, and have hole over 1,000 feet deep drilled and well finished in four days.

We have said that the stratified rocks were arranged around two centers of chystalline rocks, viz: the Black Hills, which may be compared to the horn of a saddle, and the other the Sioux Falls granite area, which runs westward from the wider granite area of Minnesota, which may be conceived to form the back and ridge of the saddle. This ridge, which may be looked upon as a buried mountain range, disappears under the later strata near Mitchell, but is traceable in wells to the vicinity of Chamberlain and will doubtless eventually be found extending nearly to the Black Hills.

Upon this saddle-like sub-stratum of granite rocks the Cambrian, Silurian, Carboniferous, Jura-Trias and Cretaceous rocks have been laid like blankets, declining to the north and the south. Those preceding the Cretaceous have been formed round the "horn," but have not reached more than half way to the east end of the state; they are exposed only around the Hills, and, as before stated, are to be described by another.

GEOLOGICAL FORMATIONS.

Crystalline or Algonkian. Before taking up the sedimentary rocks we spend a few words upon the granite of the eastern part of the state.

Near Big Stone Lake, in Minnesota, extensive quarries of granite are worked. The stone has been pronounced equal and even superior to New England granite for ornamental and building purposes. The granite extends across the line at Big Stone City, and there is an outcrop five to eight miles southwest of that place. The rock rises several feet above the general surface and there is no reason why it should not be quarried, except its greater distance from a railroad. A plant is already in operation at Aberdeen for working and polishing the Minnesota rock.

The Sioux Falls granite or quartzite, named from its prominent occurrence at that point, is a younger but, if possible, a more durable rock. The outcrops of this rock are scattered over a rudely triangular area extending to the altitude of Dell Rapids on the east line of the state, and westward to a point a few miles southeast of Mitchell. In this area there are probably three or four miles of naturally exposed surface, mostly in the valleys of streams. Such exposures have been quite generally worked for local use, but nowhere for exportation except at Sioux Falls, at east Sioux Falls, where one of the largest quarries is located, at Jasper, Dell Rapids and Spencer. The rock is very hard, strong and of a light cheerful color, sometimes of a mottled gray, but usually of different shades of pink or light purple. It is commonly fine-grained, breaks quite evenly, not only with the planes of stratification but also in other directions. It is susceptible of fine polish and is much sought for ornamental and building purposes.

Recent reports announce that an extensive outcrop of another crysalline rock has been found near Sioux Falls. It is very handsome diorite of medium grain, black and white. It is susceptible of fine and durable polish and promises to become a valuable stone.

CRETACEOUS ROCKS.

These cover nearly the whole state. Some would say that at one time they covered the Black Hills completely, and all agree that they at one time covered the whole of the eastern end of the state. At present they cover about nine-tenths of the state, though they are in turn more than half covered with the Tertiary and Quaternary formations. In their thickest development they may attain 5,000 feet or more near the Black Hills. Beginning with the oldest or lowest formation, the Cretaceous includes (1) the Dakota, (2) Colorado, (3) Montana, and (4) Laramie.

DAKOTA CRETACEOUS.

This group is named not from our state but from Dakota City, where it was first studied, which was then a pioneer town of Missouri Territory. The Dakota includes, beginning at the bottom (1) 200 to 300 feet of buff and gray sandstone, prominent at the west end of the state (Lakota), (2) a gray limestone, thirty feet thick, locally developed near Hot Springs (Minnewasta), (3) a formation consisting mostly of shales of various colors thirty to 100 feet, (Fuson), (4) a massive buff sandstone which usually forms the crest of the "Hogback" around the Hills, thirty-five to 100 feet thick (Dakota proper).

Of the valuable quarries and deposits of fire clay in this formation I leave for others to speak. It is more in order for me to dwell on a natural product furnished by this formation, which easily outranks in utility if not in nominal money value, any other natural resource of the state. Though its development is not called mining, it employes much machinery, involves much engineering, and employes some hundreds of men most of the time. Unlike most mining the product does not have to be brought laboriously to the surface but comes without effort when once set free. No, it is not petroleum or gas but a much more beneficial element—water.

Notice the position and relations of this Dakota formation. It underlies four-fifths of the state and has similar relations to the great plains generally from Canada to Texas. It is overlaid by thick, impervious clays of succeeding formations. Its western edge lies from 3,500 to 6,000 feet high on the eastern flank of the crest of the continent and around all the mountains lying east of that range, like the Black Hills. Here the water enters from the rainfall directly, from the seepage of streams which traverse its edge, and from the other porous formations which communicate with it below the surface either by faults or contact planes. The eastern edge, which lies only 1,000 to 1,200 feet above the sea, is comparatively closely sealed up by the deep covering of Cretaceous clays and glacial clays, although there are notable springs which show themselves at several points along the James and Missouri Rivers, which are doubtless outlets from this deposit. Moreover, the erosion of the glacial period and of more recent streams have so lowered the surface that one-fourth to one-third of our state may obtain flowing wells from this source and still other portions may obtain inexhaustible pump wells with water near the surface.

It also, no doubt, has large quantities of water stored within it, much of it possible at altitudes so high that it might keep up the supply for some time even if rain and river should cease.

More than 2,000 wells are now flowing in the state and are being increased by about 300 a year. They may be very roughly estimated to furnish over 70,000 gallons a minute, which would probably be about ten times the spring-time size of the Cheyenne River at Edgemont. Most of these wells are small, many an inch and a quarter in diameter, and it is now generally recognized that such wells are not only cheaper but more convenient, more serviceable and longer lived, that the large wells, such as were made several years ago. Most of the large wells have shown a steady decline, due probably to the fact that they delivered the water more rapidly than it can gather to them from the water-bearing rocks. In some places they have fallen off in flow and pressure one-quarter to one-third. In some narrow areas wells have ceased to flow, apparently from local exhaustion of water. But on the other hand, wells have been flowing nearly twenty years and still have pressures of sixty to eighty pounds to the square inch. Wells have been used for nearly that length of time for power, running electric lights, flour mills, etc., and are still in use.

In several of the wells natural gas forms an important ingredient. This is true particularly along the Missouri River from Lyman county to the north line of the state. The city of Pierre from one or two wells is abundantly supplied for lighting and for power for city purposes, and to a considerable degree for heating. Three wells in Sully county, one in Walworth and one in Campbell, in fact all which have been opened along this line, furnish gas in similar quantities. It seems not unlikely that these wells lie in the eastern border of a gas region extending possibly as far west as Meade county. It seems to be derived mainly from the same strata which furnish the water. It may possibly enter the Dakota formation from the Carboniferous underneath and may be originally derived from extensive beds of carbonaceous matter deposited along the eastern shore of the Carboniferous sea.

Lignite is found frequently in drilling wells in thin strata, but so flooded with water that no attempt has been made to obtain the product. Thin layers, twelve to thirty-six inches in thickness have been found locally developed near Ponca, Nebraska, and Sioux City, and also around the Black Hills. Petrified wood, though not of a quality suitable for ornamental purposes, is found in considerable quantities around the Black Hills.

COLORADO CRETACEOUS.

This is named from its prominence in eastern Colorado and includes a series of shales with local developments of sandstone and limestone, estimated by Mr. Darton to be from 1,450 to 1,700 feet thick around the Black Hills, and it is from 200 to 400 feet thick in the eastern end of the state. This section is commonly spoken of as the Benton from its great development near Ft. Benton on the upper Missouri. The Colorado also includes about 200 feet of chalk and calcareous shale which Dr. Hayden called the Niobrara. It is conspicuous along the Missouri River from St. Helena, Nebraska, to the great bend above Chamberlain, because of its whiteness when weathered. It is, however, often overlooked when unweathered because of its grayish tint resembling the shales above and below it. The Colorado formation contains two or three minor horizons carrying water and supplying artesian wells in the eastern part of the state, but they need not be especially distinguished from those of the Dakota.

The chalk has a very small economic value as building stone, for which it may be profitably used if carefully selected.

Its much more important use is for the manufacture of Portland cement. Its fine grain, porous structure, homogeneous character and easy grinding make it admirably adapted for mixing with clay for making a superior grade of cement. This is being extensively used for buildings and sidewalks throughout the state. Its chief factory is at Yankton, but scores of such plants might be advantageously placed along the Missouri River and around the Black Hills if there were sufficient demand.

MONTANA CRETACEOUS.

This is composed mainly of the Pierre shales named from Ft. Pierre, which are dark colored and often becoming plastic clay when wet. They are about 1,200 feet thick near the Black Hills, 300 to 400 in the eastern part of the state. They constitute the most extensive stratum of the Cretaceous, covering at least nine-tenths of the state. This is the "gumbo" of the Trans-Missouri region and constitutes probably nine-tenths of the substance of the glacial clays east of the Missouri. Hence it is a dominant element in the formation of soils over much of the state. It is rich in mineral salts favorable for grains and grasses. The prairie grasses growing upon it are noted for their nutritive and fattening qualities. Moreover, its impermiable character holds the limited rainfall near the surface and promotes rapid growth in the rainy season. Afterward it dries quickly and completely and preserves the grass as a natural hay, nutritious as grain.

Its plastic character when wet promotes its rapid erosion and the frequent occurrence of land slides which have an important effect on the topography wherever it is found. It carries little or no water, and if present of poor quality.

The Montana also includes the Fox Hills formation, 150 to 300 feet of shales and sandstone overlying the Pierre. It may possibly be a local development in the later Pierre. It caps the eastern part of the dividing plateaus between the Cheyenne and Moreau Rivers and also between the latter and the Grand. Its sandy character forms a natural mulching for the regions where it extends. Growth of grasses and crops extends over a longer period, and they are not subject to such extreme draught as upon tbe "gumbo". Springs are not infrequent, because of its attractive fossils it is often a rich field for the collector.

LARAMIE CRETACEOUS.

This, in our state, is represented by perhaps 2,500 feet of sandstone, shales, loams and clays interstratified. It is a fresh water formation unlike all preceding, which were marine. It was formed by streams, marshes and lakes. It is probably thickest in the northwest corner of the state, thins rapidly to the south and more slowly to the east. Its ragged edge extends nearly to the Black Hills on the south and across the Missouri River along the northern line of the state where it appears in conspicuous buttes.

For soil making it combines the qualities of preceding formations. It frequently exhibits fine springs. It contains especially in its upper portions, thick deposits of sandstone which in time will be very valuable for local buildings.

Undoubtedly the most valuable product of the Laramie is lignite. It has already attained prominence as a commercial product in North Dakota. There it is found in thicker beds and nearer lines of transportation, but beds five to fifteen feet thick are not uncommon in the vicinity of the Short Pine Hills, Cave Hills and Slim Buttes, and workable beds may be found north of a line extending from near the south end of Slim Buttes to the head of Fire Steel Creek, in northwestern Dewey county, and thence northeast to where Oak creek crosses the north line of the state. This includes an area within our borders of about 5,000 square miles.

Lignite differs from coal in containing a larger amount of water, which by evaporating causes it to slack. This interferes with its convenient use as a fuel. In Germany it is extensively formed into briquettes or small blocks which form a superior domestic fuel.

We look upon the Pierre as the most hopeful source of petroleum if such be found within our borders. We are led to this by the deposits in our neighboring state of Wyoming and by the fact that little or no traces of oil have been found in the drilling of the numerous wells in the eastern part of the state, several of which have gone down to crystalline rock. It must not be assumed, however, that we have sufficient evidence to arrive at any confident conclusion in this matter.

The Cretaceous was a time when reptiles ruled the world. Gigantic and strange forms swarmed upon the sea and land, and were even given wings to navigate the air. During Colorado and Montana times the forms of life were largely marine. In the Laramie huge land forms became numerous. In our views we indicate some of these.

TERTIARY FORMATIONS.

These include light colored marls, sandstones and clays which are so conspicuous in the White River Bad Lands. They are divided into the so-called White River beds, 800 to 900 feet thick in the higher points around Pine Ridge and thinning out in all directions more slowly to the east. There is also a patch in the vicinity of the Short Pine Hills and Slim Buttes. Over these lie generally, and thicker toward the east, 300 to 400 feet of loams and marls with mortar-like sandstones. These extend east of the Missouri River in the southern portion of the state in the more elevated points, like the Bijou Hills and Wessington Hills.

The peculiar erosion of these beds cause the noted White River Bad Lands of which we show characteristic views. The deposits are all of fresh water origin, the work of rivers and lakes by which the weathering of the mountains on the west were spread out in extensive sheets upon the plains on the east.

In the Tertiary times, reptiles had passed and mammals began to have their day. Nature at that time made some strange types which seem to have proved unfitting to continue, but others have by transformation lived on to the present and are now the esteemed and useful servants of man— the horse here deserves most prominent mention.

Of economic effects of these formations, we may briefly mention natural shelters for stock, frequent springs, and contributions to curiosity shops in the way of fossils, some ornamental stones are found in considerable quantities—satin spar, moss agate, and blue chalcedony or sapphirine.

Fullers' earth and volcanic ash abound and will in time be counted valuable.

QUARTERNARY FORMATIONS.

These comprise the unconsolidated deposits which lie upon the surface of other formations like a blanket and which are frequently spoken of as drift. The eastern half of the state, east of the Missouri River, is almost completely covered with a blanket of till or boulder clay, ten to 200 feet in thicknesss, lying upon highlands and lowlands alike. Associated with it are belts of stony hills or moraines, lake beds and ancient channels are frequent features. Here are included also the numerous terraces, some of them 300 or 400 feet above the present stream, and sometimes several miles in width, covering with sand and loam which come in to modify the effects of the formations hitherto discussed.

These terraces are particularly prominent along the western tributaries of the Missouri, but are also conspicuous on that stream and along channels now vacated but occupied during the glacial period.

The marked effects of the glacial period upon the geology of our state we need not dwell upon, but turn our attention more to the economic results which many may overlook.

We sometimes become impressed by the great expense necessary to prepare the natural surface for the proper location of manufacturnig plants, irrigation projects, or the building of cities. The work of the glacial period, especially in the eastern half of the state, can scarcely be overestimated from an economical standpoint. By it the surface was smoothed and beautifully graded for agricultural purposes, natural basins were formed for the retention of rainfall, thus giving an object lesson to man for the further improvement of the region, extensive deposits of sand and gravel were formed, the components of various formations were intermingled and ground together to form a rich sub-soil, picturesque lakes and pleasing elevations were formed for pleasure resorts, and extensive terraces conveniently located along prominent streams seem naturally prepared for suitable locations of cities and towns.

It scarcely need be stated that no traces of precious metals have been found outside of the Hills. While in California and other localities gold has been found in Mesozoic and Tertiary strata, it should be remembered that it has always been in connection with marked disturbance of the earth's crust with the formation of veins and the outflow of igneous rocks. No such disturbance has yet been noted in our borders. Strata have been somewhat tilted in the Slim Buttes and profound crevices have been formed in the tertiary of the Bad Lands and filled with sand, gypsum and quartz, but these have evidently failed to reach to the deep-seated waters which are the usual vehicles of precious metals.

The finding of gold has been reported from several localities but it has invariably been found to rest upon very superficial evidence. At a few points in the eastern part of the state very minute quantities have been found in the glacial drift, which may be reasonably referred to the region of the Lake of the Woods as their probable origin. The most clear case of this sort was at Gary several years ago.

This is the story of the rocks of our state outside the Hills so far as has yet been interpreted. It is full of promise. Nature has done her part, probably better than has been sometimes thought. Wherein our circumstances are novel or peculiar a hint is given us of the perculiar destiny to which a kind Providence has called us. The secret of commercial and social success in our commonwealth is to learn the truth concerning our resources and the best methods of utilizing them. Let us go on in an honest, generous spirit to make the most of them patiently and hopefully, and to welcome and encourage all who may cast in their lot with us.

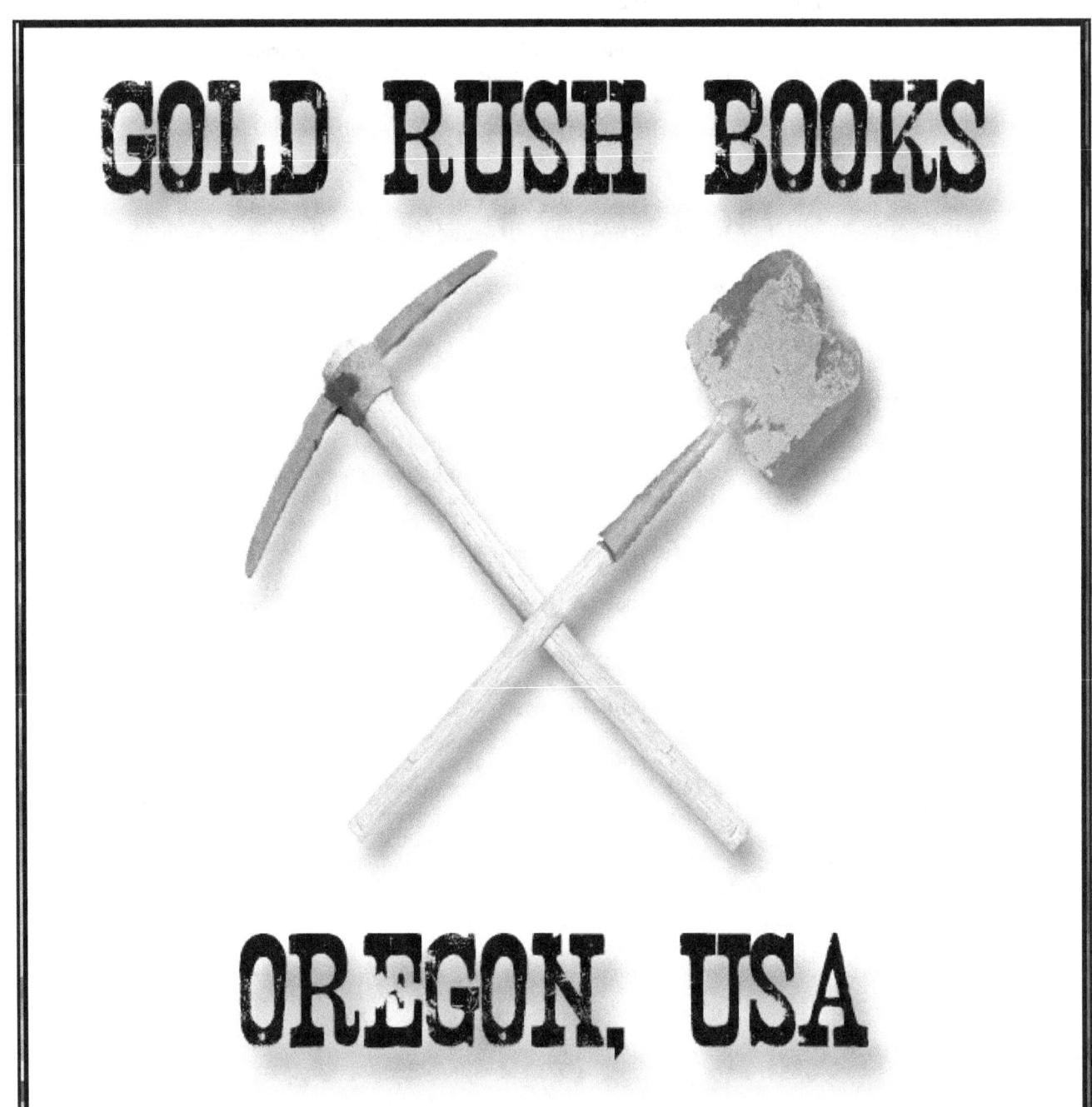

GOLD RUSH BOOKS

OREGON, USA

www.GoldMiningBooks.com

Books On Mining

Visit: www.goldminingbooks.com to order your copies or ask your favorite book seller to offer them.

Mining Books by Kerby Jackson

Gold Dust: Stories From Oregon's Mining Years - Oregon mining historian and prospector, Kerby Jackson, brings you a treasure trove of seventeen stories on Southern Oregon's rich history of gold prospecting, the prospectors and their discoveries, and the breathtaking areas they settled in and made homes. 5" X 8", 98 ppgs. Retail Price: $11.99

The Golden Trail: More Stories From Oregon's Mining Years - In his follow-up to "Gold Dust: Stories of Oregon's Mining Years", this time around, Jackson brings us twelve tales from Oregon's Gold Rush, including the story about the first gold strike on Canyon Creek in Grant County, about the old timers who found gold by the pail full at the Victor Mine near Galice, how Iradel Bray discovered a rich ledge of gold on the Coquille River during the height of the Rogue River War, a tale of two elderly miners on the hunt for a lost mine in the Cascade Mountains, details about the discovery of the famous Armstrong Nugget and others. 5" X 8", 70 ppgs. Retail Price: $10.99

Oregon Mining Books

Geology and Mineral Resources of Josephine County, Oregon - Unavailable since the 1970's, this important publication was originally compiled by the Oregon Department of Geology and Mineral Industries and includes important details on the economic geology and mineral resources of this important mining area in South Western Oregon. Included are notes on the history, geology and development of important mines, as well as insights into the mining of gold, copper, nickel, limestone, chromium and other minerals found in large quantities in Josephine County, Oregon. 8.5" X 11", 54 ppgs. Retail Price: $9.99

Mines and Prospects of the Mount Reuben Mining District - Unavailable since 1947, this important publication was originally compiled by geologist Elton Youngberg of the Oregon Department of Geology and Mineral Industries and includes detailed descriptions, histories and the geology of the Mount Reuben Mining District in Josephine County, Oregon. Included are notes on the history, geology, development and assay statistics, as well as underground maps of all the major mines and prospects in the vicinity of this much neglected mining district. 8.5" X 11", 48 ppgs. Retail Price: $9.99

The Granite Mining District - Notes on the history, geology and development of important mines in the well known Granite Mining District which is located in Grant County, Oregon. Some of the mines discussed include the Ajax, Blue Ribbon, Buffalo, Continental, Cougar-Independence, Magnolia, New York, Standard and the Tillicum. Also included are many rare maps pertaining to the mines in the area. 8.5" X 11", 48 ppgs. Retail Price: $9.99

Ore Deposits of the Takilma and Waldo Mining Districts of Josephine County, Oregon - The Waldo and Takilma mining districts are most notable for the fact that the earliest large scale mining of placer gold and copper in Oregon took place in these two areas. Included are details about some of the earliest large gold mines in the state such as the Llano de Oro, High Gravel, Cameron, Platerica, Deep Gravel and others, as well as copper mines such as the famous Queen of Bronze mine, the Waldo, Lily and Cowboy mines. This volume also includes six maps and 20 original illustrations. 8.5" X 11", 74 ppgs. Retail Price: $9.99

Metal Mines of Douglas, Coos and Curry Counties, Oregon - Oregon mining historian Kerby Jackson introduces us to a classic work on Oregon's mining history in this important re-issue of Bulletin 14C Volume 1, otherwise known as the Douglas, Coos & Curry Counties, Oregon Metal Mines Handbook. Unavailable since 1940, this important publication was originally compiled by the Oregon Department of Geology and Mineral Industries includes detailed descriptions, histories and the geology of over 250 metallic mineral mines and prospects in this rugged area of South West Oregon. 8.5" X 11", 158 ppgs. Retail Price: $19.99

Metal Mines of Jackson County, Oregon - Unavailable since 1943, this important publication was originally compiled by the Oregon Department of Geology and Mineral Industries includes detailed descriptions, histories and the geology of over 450 metallic mineral mines and prospects in Jackson County, Oregon. Included are such famous gold mining areas as Gold Hill, Jacksonville, Sterling and the Upper Applegate. 8.5" X 11", 220 ppgs. Retail Price: $24.99

Metal Mines of Josephine County, Oregon - Oregon mining historian Kerby Jackson introduces us to a classic work on Oregon's mining history in this important re-issue of Bulletin 14C, otherwise known as the Josephine County, Oregon Metal Mines Handbook. Unavailable since 1952, this important publication was originally compiled by the Oregon Department of Geology and Mineral Industries includes detailed descriptions, histories and the geology of over 500 metallic mineral mines and prospects in Josephine County, Oregon. 8.5" X 11", 250 ppgs. Retail Price: $24.99

Metal Mines of North East Oregon - Oregon mining historian Kerby Jackson introduces us to a classic work on Oregon's mining history in this important re-issue of Bulletin 14A and 14B, otherwise known as the North East Oregon Metal Mines Handbook. Unavailable since 1941, this important publication was originally compiled by the Oregon Department of Geology and Mineral Industries and includes detailed descriptions, histories and the geology of over 750 metallic mineral mines and prospects in North Eastern Oregon. 8.5" X 11", 310 ppgs. Retail Price: $29.99

Metal Mines of North West Oregon - Oregon mining historian Kerby Jackson introduces us to a classic work on Oregon's mining history in this important re-issue of Bulletin 14D, otherwise known as the North West Oregon Metal Mines Handbook. Unavailable since 1951, this important publication was originally compiled by the Oregon Department of Geology and Mineral Industries and includes detailed descriptions, histories and the geology of over 250 metallic mineral mines and prospects in North Western Oregon. 8.5" X 11", 182 ppgs. Retail Price: $19.99

Mines and Prospects of Oregon - Mining historian Kerby Jackson introduces us to a classic mining work by the Oregon Bureau of Mines in this important re-issue of The Handbook of Mines and Prospects of Oregon. Unavailable since 1916, this publication includes important insights into hundreds of gold, silver, copper, coal, limestone and other mines that operated in the State of Oregon around the turn of the 19th Century. Included are not only geological details on early mines throughout Oregon, but also insights into their history, production, locations and in some cases, also included are rare maps of their underground workings. 8.5" X 11", 314 ppgs. Retail Price: $24.99

Lode Gold of the Klamath Mountains of Northern California and South West Oregon
(See California Mining Books)

Mineral Resources of South West Oregon - Unavailable since 1914, this publication includes important insights into dozens of mines that once operated in South West Oregon, including the famous gold fields of Josephine and Jackson Counties, as well as the Coal Mines of Coos County. Included are not only geological details on early mines throughout South West Oregon, but also insights into their history, production and locations. 8.5" X 11", 154 ppgs. Retail Price: $11.99

Chromite Mining in The Klamath Mountains of California and Oregon
(See California Mining Books)

Southern Oregon Mineral Wealth - Unavailable since 1904, this rare publication provides a unique snapshot into the mines that were operating in the area at the time. Included are not only geological details on early mines throughout South West Oregon, but also insights into their history, production and locations. Some of the mining areas include Grave Creek, Greenback, Wolf Creek, Jump Off Joe Creek, Granite Hill, Galice, Mount Reuben, Gold Hill, Galls Creek, Kane Creek, Sardine Creek, Birdseye Creek, Evans Creek, Foots Creek, Jacksonville, Ashland, the Applegate River, Waldo, Kerby and the Illinois River, Althouse and Sucker Creek, as well as insights into local copper mining and other topics. 8.5" X 11", 64 ppgs. Retail Price: $8.99

Geology and Ore Deposits of the Takilma and Waldo Mining Districts - Unavailable since the 1933, this publication was originally compiled by the United States Geological Survey and includes details on gold and copper mining in the Takilma and Waldo Districts of Josephine County, Oregon. The Waldo and Takilma mining districts are most notable for the fact that the earliest large scale mining of placer gold and copper in Oregon took place in these two areas. Included in this report are details about some of the earliest large gold mines in the state such as the Llano de Oro, High Gravel, Cameron, Platerica, Deep Gravel and others, as well as copper mines such as the famous Queen of Bronze mine, the Waldo, Lily and Cowboy mines. In addition to geological examinations, insights are also provided into the production, day to day operations and early histories of these mines, as well as calculations of known mineral reserves in the area. This volume also includes six maps and 20 original illustrations. 8.5" X 11", 74 ppgs. Retail Price: $9.99

Gold Mines of Oregon - Oregon mining historian Kerby Jackson introduces us to a classic work on Oregon's mining history in this important re-issue of Bulletin 61, otherwise known as "Gold and Silver In Oregon". Unavailable since 1968, this important publication was originally compiled by geologists Howard C. Brooks and Len Ramp of the Oregon Department of Geology and Mineral Industries and includes detailed descriptions, histories and the geology of over 450 gold mines Oregon. Included are notes on the history, geology and gold production statistics of all the major mining areas in Oregon including the Klamath Mountains, the Blue Mountains and the North Cascades. While gold is where you find it, as every miner knows, the path to success is to prospect for gold where it was previously found. **8.5" X 11", 344 ppgs. Retail Price: $24.99**

Mines and Mineral Resources of Curry County Oregon - Originally published in 1916, this important publication on Oregon Mining has not been available for nearly a century. Included are rare insights into the history, production and locations of dozens of gold mines in Curry County, Oregon, as well as detailed information on important Oregon mining districts in that area such as those at Agness, Bald Face Creek, Mule Creek, Boulder Creek, China Diggings, Collier Creek, Elk River, Gold Beach, Rock Creek, Sixes River and elsewhere. Particular attention is especially paid to the famous beach gold deposits of this portion of the Oregon Coast. **8.5" X 11", 140 ppgs. Retail Price: $11.99**

Chromite Mining in South West Oregon - Originally published in 1961, this important publication on Oregon Mining has not been available for nearly a century. Included are rare insights into the history, production and locations of nearly 300 chromite mines in South Western Oregon. **8.5" X 11", 184 ppgs. Retail Price: $14.99**

Mineral Resources of Douglas County Oregon - Originally published in 1972, this important publication on Oregon Mining has not been available for nearly forty years. Included are rare insights into the geology, history, production and locations of numerous gold mines and other mining properties in Douglas County, Oregon. **8.5" X 11", 124 ppgs. Retail Price: $11.99**

Mineral Resources of Coos County Oregon - Originally published in 1972, this important publication on Oregon Mining has not been available for nearly forty years. Included are rare insights into the geology, history, production and locations of numerous gold mines and other mining properties in Coos County, Oregon. **8.5" X 11", 100 ppgs. Retail Price: $11.99**

Mineral Resources of Lane County Oregon - Originally published in 1938, this important publication on Oregon Mining has not been available for nearly seventy five years. Included are extremely rare insights into the geology and mines of Lane County, Oregon, in particular in the Bohemia, Blue River, Oakridge, Black Butte and Winberry Mining Districts. **8.5" X 11", 82 ppgs. Retail Price: $9.99**

Mineral Resources of the Upper Chetco River of Oregon: Including the Kalmiopsis Wilderness - Originally published in 1975, this important publication on Oregon Mining has not been available for nearly forty years. Withdrawn under the 1872 Mining Act since 1984, real insight into the minerals resources and mines of the Upper Chetco River has long been unavailable due to the remoteness of the area. Despite this, the decades of battle between property owners and environmental extremists over the last private mining inholding in the area has continued to pique the interest of those interested in mining and other forms of natural resource use. Gold mining began in the area in the 1850's and has a rich history in this geographic area, even if the facts surrounding it are little known. Included are twenty two rare photographs, as well as insights into the Becca and Morning Mine, the Emmly Mine (also known as Emily Camp), the Frazier Mine, the Golden Dream or Higgins Mine, Hustis Mine, Peck Mine and others. **8.5" X 11", 64 ppgs. Retail Price: $8.99**

Gold Dredging in Oregon - Originally published in 1939, this important publication on Oregon Mining has not been available for nearly seventy five years. Included are extremely rare insights into the history and day to day operations of the dragline and bucketline gold dredges that once worked the placer gold fields of South West and North East Oregon in decades gone by. Also included are details into the areas that were worked by gold dredges in Josephine, Jackson, Baker and Grant counties, as well as the economic factors that impacted this mining method. This volume also offers a unique look into the values of river bottom land in relation to both farming and mining, in how farm lands were mined, re-soiled and reclamated after the dredges worked them. Featured are hard to find maps of the gold dredge fields, as well as rare photographs from a bygone era. **8.5" X 11", 86 ppgs. Retail Price: $8.99**

Quick Silver Mining in Oregon - Originally published in 1963, this important publication on Oregon Mining has not been available for over fifty years. This publication includes details into the history and production of Elemental Mercury or Quicksilver in the State of Oregon. **8.5" X 11", 238 ppgs. Retail Price: $15.99**

Mines of the Greenhorn Mining District of Grant County Oregon - Originally published in 1948, this important publication on Oregon Mining has not been available for over sixty five years. In this publication are rare insights into the mines of the famous Greenhorn Mining District of Grant County, Oregon, especially the famous Morning Mine. Also included are details on the Tempest, Tiger, Bi-Metallic, Windsor, Psyche, Big Johnny, Snow Creek, Banzette and Paramount Mines, as well as prospects in the vicinities in the famous mining areas of Mormon Basin, Vinegar Basin and Desolation Creek. Included are hard to find mine maps and dozens of rare photographs from the bygone era of Grant County's rich mining history. **8.5" X 11", 72 ppgs. Retail Price: $9.99**

Geology of the Wallowa Mountains of Oregon: Part I (Volume 1) - Originally published in 1938, this important publication on Oregon Mining has not been available for nearly seventy five years. Included are details on the geology of this unique portion of North Eastern Oregon. This is the first part of a two book series on the area. Accompanying the text are rare photographs and historic maps.**8.5" X 11", 92 ppgs. Retail Price: $9.99**

Geology of the Wallowa Mountains of Oregon: Part II (Volume 2) - Originally published in 1938, this important publication on Oregon Mining has not been available for nearly seventy five years. Included are details on the geology of this unique portion of North Eastern Oregon. This is the first part of a two book series on the area. Accompanying the text are rare photographs and historic maps.**8.5" X 11", 94 ppgs. Retail Price: $9.99**

Field Identification of Minerals For Oregon Prospectors - Originally published in 1940, this important publication on Oregon Mining has not been available for nearly seventy five years. Included in this volume is an easy system for testing and identifying a wide range of minerals that might be found by prospectors, geologists and rockhounds in the State of Oregon, as well as in other locales. Topics include how to put together your own field testing kit and how to conduct rudimentary tests in the field. This volume is written in a clear and concise way to make it useful even for beginners. **8.5" X 11", 158 ppgs. Retail Price: $14.99**

The Bohemia Mining District of Oregon - Originally published in 1900, this important publication on Oregon Mining has not been available for over a century. Included in this volume are important insights into the famous Bohemia Mining District of Oregon, including the histories and locations of important gold mines in the area such as the Ophir Mine, Clarence, Acturas, Peek-a-boo, White Swan, Combination Mine, the Musick Mine, The California, White Ghost, The Mystery, Wall Street, Vesuvius, Story, Lizzie Bullock, Delta, Elsie Dora, Golden Slipper, Broadway, Champion Mine, Knott, Noonday, Helena, White Wings, Riverside and others. Also included are notes on the nearby Blue River Mining District. **8.5" X 11", 58 ppgs. Retail Price: $9.99**

The Gold Fields of Eastern Oregon - Unavailable since 1900, this publication was originally compiled by the Baker City Chamber of Commerce Offering important insights into the gold mining history of Eastern Oregon, "The Gold Fields of Eastern Oregon" sheds a rare light on many of the gold mines that were operating at the turn of the 19th Century in Baker County and Grant County in North Eastern Oregon. Some of the areas featured include the Cable Cove District, Baisely-Elhorn, Granite, Red Boy, Bonanza, Susanville, Sparta, Virtue, Vaughn, Sumpter, Burnt River, Rye Valley and other mining districts. Included is basic information on not only many gold mines that are well known to those interested in Eastern Oregon mining history, but also many mines and prospects which have been mostly lost to the passage of time. Accompanying are numerous rare photos **8.5" X 11", 78 ppgs. Retail Price: $10.99**

Gold Mining in Eastern Oregon - Originally published in 1938, this important publication on Oregon Mining has not been available for over a century. Included in this volume are important insights into the famous mining districts of Eastern Oregon during the late 1930's. Particular attention is given to those gold mines with milling and concentrating facilities in the Greenhorn, Red Boy, Alamo, Bonanza, Granite, Cable Cove, Cracker Creek, Virtue, Keating, Medical Springs, Sanger, Sparta, Chicken Creek, Mormon Basin, Connor Creek, Cornucopia and the Bull Run Mining Districts. Some of the mines featured include the Ben Harrison, North Pole-Columbia, Highland Maxwell, Baisley-Elkhorn, White Swan, Balm Creek, Twin Baby, Gem of Sparta, New Deal, Gleason, Gifford-Johnson, Cornucopia, Record, Bull Run, Orion and others. Of particular interest are the mill flow sheets and descriptions of milling operations of these mines. **8.5" X 11", 68 ppgs. Retail Price: $8.99**

The Gold Belt of the Blue Mountains of Oregon - Originally published in 1901, this important publication on Oregon Mining has not been available for over a century. Included in this volume are rare insights into the gold deposits of the Blue Mountains of North East Oregon, including the history of their early discovery and early production. Extensive details are offered on this important mining area's mineralogy and economic geology, as well as insights into nearby gold placers, silver deposits and copper deposits. Featured are the Elkhorn and Rock Creek mining districts, the Pocahontas district, Auburn and Minersville districts, Sumpter and Cracker Creek, Cable Cove, the Camp Carson district, Granite, Alamo, Greenhorn, Robinsonville, the Upper Burnt River Valley and Bonanza districts, Susanville, Quartzburg, Canyon Creek, Virtue, the Copper Butte district, the North Powder River, Sparta, Eagle Creek, Cornucopia, Pine Creek, Lower Powder River, the Upper Snake River Canyon, Rye Valley, Lower Burnt River Valley, Mormon Basin, the Malheur and Clarks Creek districts, Sutton Creek and others. Of particular interest are important details on numerous gold mines and prospects in these mining districts, including their locations, histories, geology and other important information, as well as information on silver, copper and fire opal deposits. **8.5" X 11", 250 ppgs. Retail Price: $24.99**

Mining in the Cascades Range of Oregon - Originally published in 1938, this important publication on Oregon Mining has not been available for over seventy five years. Included in this volume are rare insights into the gold mines and other types of metal mines in the Cascades Mountain Range of Oregon. Some of the important mining areas covered include the famous Bohemia Mining District, the North Santiam Mining District, Quartzville Mining District, Blue River Mining District, Fall Creek Mining District, Oakridge District, Zinc District, Buzzard-Al Sarena District, Grand Cove, Climax District and Barron Mining District. Of particular interest are important details on over 100 mines and prospects in these mining districts, including their locations, histories, geology and other important information. 8.5" X 11", 170 ppgs. **Retail Price: $14.99**

Beach Gold Placers of the Oregon Coast - Originally published in 1934, this important publication on Oregon Mining has not been available for over 80 years. Included in this volume are rare insights into the beach gold deposits of the State of Oregon, including their locations, occurance, composition and geology. Of particular interest is information on placer platinum in Oregon's rich beach deposits. Also included are the locations and other information on some famous Oregon beach mines, including the Pioneer, Eagle, Chickamin, Iowa and beach placer mines north of the mouth of the Rogue River. **8.5" X 11", 60 ppgs. Retail Price: $8.99**

Mineralogical Composition of the Sands of the Oregon Coast: From Coos Bay to the Columbia - Published in 1945, he text features hard to find information on the composition of the gold bearing black sands of the South West Oregon Coast, offering a unique insight to prospectors in search of Oregon's legendary beach gold. 104 ppgs, $9.99

Manganese Mining in Oregon - First released in 1942 and now out of print, this special reprint edition of "Manganese in Oregon" was originally published by the Oregon Department of Geology and Mineral Industries. The text features hard to find information on the mining of Manganese in Oregon, including details and maps of Oregon manganese mines and prospects. 108 ppgs, 9.99

Medford Oregon As A Mining Center - Written in 1912, this hard to find publication includes valuable insights into the mining history of South West Oregon. This small book contains interesting information on the gold, copper and mining industry in Southern Oregon as it existed just prior to World War One, shedding light on some of the important mines in the area. Included are rare photographs and vintage advertising of the day. 80 ppgs, 9.99

Mineral Resources of Curry County Oregon - First released in 1977 and now out of print, this special reprint edition of "Geology, Mineral Resources and Rock Materials of Curry County, Oregon" was originally published in cooperation of Curry County, Oregon and the Oregon Department of Geology and Mineral Industries. The text features hard to find information on not only the mining of gold and other metals in Curry County, but also aggregate mining in the area. 102 ppgs, 11.99

Origin of the Gold Bearing Black Sands of the Coast of South West Oregon - First released in 1943 and now out of print, this special reprint edition of "The Origin of the Black Sands of the South West Oregon Coast" was originally published by the Oregon Department of Geology and Mineral Industries. The text features hard to find information on the origin of the gold bearing black sands of the South West Oregon Coast, offering a unique insight to prospectors in search of Oregon's legendary beach gold. 52 ppgs, 8.99

South West Oregon Mining - Leading mining historian Kerby Jackson introduces us to six classic small mining publications on the Gold Mining Industry in Southern Oregon. This small book consists of a compilation of USGS J.S. Diller's "Mines of the Riddles Quadrangle", "The Rogue River Valley Coal Fields" and "Mineral Resources of the Grants Pass Quadrangle", the Grants Pass Commercial Club's rare publication "Mining in Josephine County, Oregon" and the USGS publication "The Distribution of Placer Gold in the Sixes River, South West Oregon". Also included is F.W. Libbey's legendary article on the Southern Oregon Mining Industry, "Lest We Forget", which appeared in the publication of the Oregon State Department of Geology and Mineral Industries in the early 1960's. This compilation offers a unique perspective on mining in South West Oregon and includes considerable information on mines in Josephine, Jackson and Coos Counties. 142 ppgs, 14.99

Geology and Mineral Resources of the Gasquet Quadrangle of California-Oregon - First published in 1953, it has been unavailable for over a century and sheds important light on the geological features and mineral resources of this portion of Northern California and Southern Oregon. 80 ppgs, 9.99

Idaho Mining Books

Gold in Idaho - Unavailable since the 1940's, this publication was originally compiled by the Idaho Bureau of Mines and includes details on gold mining in Idaho. Included is not only raw data on gold production in Idaho, but also valuable insight into where gold may be found in Idaho, as well as practical information on the gold bearing rocks and other geological features that will assist those looking for placer and lode gold in the State of Idaho. This volume also includes thirteen gold maps that greatly enhance the practical usability of the information contained in this small book detailing where to find gold in Idaho. **8.5" X 11", 72 ppgs. Retail Price: $9.99**

Geology of the Couer D'Alene Mining District of Idaho - Unavailable since 1961, this publication was originally compiled by the Idaho Bureau of Mines and Geology and includes details on the mining of gold, silver and other minerals in the famous Coeur D'Alene Mining District in Northern Idaho. Included are details on the early history of the Coeur D'Alene Mining District, local tectonic settings, ore deposit features, information on the mineral belts of the Osburn Fault, as well as detailed information on the famous Bunker Hill Mine, the Dayrock Mine, Galena Mine, Lucky Friday Mine and the infamous Sunshine Mine. This volume also includes sixteen hard to find maps. **8.5" X 11", 70 ppgs. Retail Price: $9.99**

The Gold Camps and Silver Cities of Idaho - Originally published in 1963, this important publication on Idaho Mining has not been available for nearly fifty years. Included are rare insights into the history of Idaho's Gold Rush, as well as the mad craze for silver in the Idaho Panhandle. Documented in fine detail are the early mining excitements at Boise Basin, at South Boise, in the Owyhees, at Deadwood, Long Valley, Stanley Basin and Robinson Bar, at Atlanta, on the famous Boise River, Volcano, Little Smokey, Banner, Boise Ridge, Hailey, Leesburg, Lemhi, Pearl, at South Mountain, Shoup and Ulysses, Yellow Jacket and Loon Creek. The story follows with the appearance of Chinese miners at the new mining camps on the Snake River, Black Pine, Yankee Fork, Bay Horse, Clayton, Heath, Seven Devils, Gibbonsville, Vienna and Sawtooth City. Also included are special sections on the Idaho Lead and Silver mines of the late 1800's, as well as the mining discoveries of the early 1900's that paved the way for Idaho's modern mining and mineral industry. Lavishly illustrated with rare historic photos, this volume provides a one of a kind documentary into Idaho's mining history that is sure to be enjoyed by not only modern miners and prospectors who still scour the hills in search of nature's treasures, but also those enjoy history and tromping through overgrown ghost towns and long abandoned mining camps. **8.5" X 11", 186 ppgs. Retail Price: $14.99**

Ore Deposits and Mining in North Western Custer County Idaho - Unavailable since 1913, this important publication was originally published by the Us Department of the Interior and has been unavailable for a century. Included are fine details on the geology, geography, gold placers and gold and silver bearing quartz veins of the mining region of North West Custer County, Idaho. Of particular interest is a rare look at the mines and prospects of the region, including those such as the Ramshorn Mine, SkyLark, Riverview, Excelsior, Beardsley, Pacific, Hoosier, Silver Brick, Forest Rose and dozens of others in the Bay Horse Mining District. Also covered are the mines of the Yankee Fork District such as the Lucky Boy, Badger, Black, Enterprise, Charles Dickens, Morrison, Golden Sunbeam, Montana, Golden Gate and others, as well as those in the Loon Mining District. **8.5" X 11", 126 ppgs. Retail Price: $12.99**

Gold Rush To Idaho - Unavailable since 1963, this important publication was originally published by the Idaho Bureau of Mines and has been unavailable for 50 years. "Gold Rush To Idaho" revisits the earliest years of the discovery of gold in Idaho Territory and introduces us to the conditions that the pioneer gold seekers met when they blazed a trail through the wilderness of Idaho's mountains and discovered the precious yellow metal at Oro Fino and Pierce. Subsequent rushes followed at places like Elk City, Newsome, Clearwater Station, Florence, Warrens and elsewhere. Of particular interest is a rare look at the hardships that the first miners in Idaho met with during their day to day existences and their attempts to bring law and order to their mining camps. **8.5" X 11", 88 ppgs. Retail Price: $9.99**

The Geology and Mines of Northern Idaho and North Western Montana - Unavailable since 1909, this important publication was originally published by the Us Department of the Interior and has been unavailable for a century. Included are fine details on the geology and geography of the mining regions of Northern Idaho and North Western Montana. Of particular interest is a rare look at the mines and prospects of the region, including those in the Pine Creek Mining District, Lake Pend Oreille district, Troy Mining District, Sylvanite District, Cabinet Mining District, Prospect Mining District and the Missoula Valley. Some of the mines featured include the Iron Mountain, Silver Butte, Snowshoe, Grouse Mountain Mine and others. **8.5" X 11", 142 ppgs. Retail Price: $12.99**

Mining in the Alturas Quadrangle of Blaine County Idaho - Unavailable since 1922, this important publication was originally published by the Idaho Bureau of Mines and has been unavailable for ninety years. Topics include the geology, rock formations and the formation of ore deposits in this important mining area of Idaho. Of particular focus is information on the local geology, quartz veins and ore deposits of this portion of Idaho. Included are hard to find details, including the descriptions and locations of numerous gold and silver mines in the area including the Silver King, Pilgrim, Columbia, Lone Jack, Sunbeam, Pride of the West, Lucky Boy, Scotia, Atlanta, Beaver-Bidwell and others mines and prospects. **8.5" X 11", 56 ppgs. Retail Price: $8.99**

Mining in Lemhi County Idaho - Originally published in 1913, this important book on Idaho Mining has not been available to miners for over a century. Included are rare insights into hundreds of gold, silver, copper and other mines in this famous Idaho mining area. Details include the locations, geology, history, production and other facts of the mines of this region, not only gold and silver hardrock mines, but also gold placer mines, lead-silver deposits, copper mines, cobalt-nickel deposits, tungsten and tin mines . It is lavishly illustrated with hard to find photos of the period and rare mining maps. Some of the vicinities featured include the Nicholia Mining District, Spring Mountain District, Texas District, Blue Wing District, Junction District, McDevitt District, Pratt Creek, Eldorado District, Kirtley Creek, Carmen Creek, Gibbonsville, Indian Creek, Mineral Hill District, Mackinaw, Eureka District, Blackbird District, YellowJacket District, Gravel Range District, Junction District, Parker Mountain and other mining districts. 8.5" X 11", 226 ppgs. Retail Price: $19.99

Mining in Shoshone County Idaho - First published in 1923, it has been unavailable for over a century and sheds important light on the mining history of Shoshone County, Idaho. Some of the topics include the history of mining in Shoshone County, a look at the local geology and ore characteristics of lead-silver deposits, zinc deposits, copper, antimony, gold and other minerals. Also included are insights into the history, production, characteristics and locations of numerous mines in the area. 198 ppgs, 15.99

Utah Mining Books

Fluorite in Utah - Unavailable since 1954, this publication was originally compiled by the USGS, State of Utah and U.S. Atomic Energy Commission and details the mining of fluorspar, also known as fluorite in the State of Utah. Included are details on the geology and history of fluorspar (fluorite) mining in Utah, including details on where this unique gem mineral may be found in the State of Utah. 8.5" X 11", 60 ppgs. Retail Price: $8.99

The Gold Hill Mining District of Utah - First published in 1935, it has been unavailable since those days and sheds important light on the mines, history and geology of Utah's Gold Hill Mining District. Included are rare insights into this important mining area, including the locations, histories and details of numerous mines. This volume is well illustrated with geological diagrams, as well as hard to find maps of some of the most important mines in this district. 202 ppgs., 19.99

The Mines, Miners and Minerals of Utah - First published in 1896, it has been unavailable since those days and sheds important light on the early mines and miners of Pioneer Utah, as well as the minerals which they won from the earth by laborious hard physical labor and sheer determination. Included are rare insights into the early mining history of Utah, as well details on hundreds of gold, silver and copper mines. 376 ppgs., 24.99

California Mining Books

The Tertiary Gravels of the Sierra Nevada of California - Mining historian Kerby Jackson introduces us to a classic mining work by Waldemar Lindgren in this important re-issue of The Tertiary Gravels of the Sierra Nevada of California. Unavailable since 1911, this publication includes details on the gold bearing ancient river channels of the famous Sierra Nevada region of California. 8.5" X 11", 282 ppgs. Retail Price: $19.99

The Mother Lode Mining Region of California - Unavailable since 1900, this publication includes details on the gold mines of California's famous Mother Lode gold mining area. Included are details on the geology, history and important gold mines of the region, as well as insights into historic mining methods, mine timbering, mining machinery, mining bell signals and other details on how these mines operated. Also included are insights into the gold mines of the California Mother Lode that were in operation during the first sixty years of California's mining history. 8.5" X 11", 176 ppgs. Retail Price: $14.99

Lode Gold of the Klamath Mountains of Northern California and South West Oregon - Unavailable since 1971, this publication was originally compiled by Preston E. Hotz and includes details on the lode mining districts of Oregon and California's Klamath Mountains. Included are details on the geology, history and important lode mines of the French Gulch, Deadwood, Whiskeytown, Shasta, Redding, Muletown, South Fork, Old Diggings, Dog Creek (Delta), Bully Choop (Indian Creek), Harrison Gulch, Hayfork, Minersville, Trinity Center, Canyon Creek, East Fork, New River, Denny, Liberty (Black Bear), Cecilville, Callahan, Yreka, Fort Jones and Happy Camp mining districts in California, as well as the Ashland, Rogue River, Applegate, Illinois River, Takilma, Greenback, Galice, Silver Peak, Myrtle Creek and Mule Creek districts of South Western Oregon. Also included are insights into the mineralization and other characteristics of this important mining region. 8.5" X 11", 100 ppgs. Retail Price: $10.99

Mines and Mineral Resources of Shasta County, Siskiyou County, Trinity County: California - Unavailable since 1915, this publication was originally compiled by the California State Mining Bureau and includes details on the gold mines of this area of Northern California. Also included are insights into the mineralization and other characteristics of this important mining region, as well as the location of historic gold mines. 8.5" X 11", 204 ppgs. Retail Price: $19.99

Geology of the Yreka Quadrangle, Siskiyou County, California - Unavailable since 1977, this publication was originally compiled by Preston E. Hotz and includes details on the geology of the Yreka Quadrangle of Siskiyou County, California. Also included are insights into the mineralization and other characteristics of this important mining region. 8.5" X 11", 78 ppgs. Retail Price: $7.99

Mines of San Diego and Imperial Counties, California - Originally published in 1914, this important publication on California Mining has not been available for a century. This publication includes important information on the early gold mines of San Diego and Imperial County, which were some of the first gold fields mined in California by early Spanish and Mexican miners before the 49ers came on the scene. Included are not only details on early mining methods in the area, production statistics and geological information, but also the location of the early gold mines that helped make California "The Golden State". Also included are details on the mining of other minerals such as silver, lead, zinc, manganese, tungsten, vanadium, asbestos, barite, borax, cement, clay, dolomite, fluospar, gem stones, graphite, marble, salines, petroleum, stronium, talc and others. 8.5" X 11", 116 ppgs. Retail Price: $12.99

Mines of Sierra County, California - Unavailable since 1920, this publication was originally compiled by the California State Mining Bureau and includes details on the gold mines of Sierra County, California. Also included are insights into the mineralization and other characteristics of this important mining region, as well as the location of historic gold mines. 8.5" X 11", 156 ppgs. Retail Price: $19.99

Mines of Plumas County, California - Unavailable since 1918, this publication was originally compiled by the California State Mining Bureau and includes details on the gold mines of Plumas County, California. Also included are insights into the mineralization and other characteristics of this important mining region, as well as the location of historic gold mines. 8.5" X 11", 200 ppgs. Retail Price: $19.99

Mines of El Dorado, Placer, Sacramento and Yuba Counties, California - Originally published in 1917, this important publication on California Mining has not been available for nearly a century. This publication includes important information on the early gold mines of El Dorado County, Placer County, Sacramento County and Yuba County, which were some of the first gold fields mined by the Forty-Niners during the California Gold Rush. Included are not only details on early mining methods in the area, production statistics and geological information, but also the location of the early gold mines that helped make California "The Golden State". Also included are insights into the early mining of chrome, copper and other minerals in this important mining area. 8.5" X 11", 204 ppgs. Retail Price: $19.99

Mines of Los Angeles, Orange and Riverside Counties, California - Originally published in 1917, this important publication on California Mining has not been available for nearly a century. This publication includes important information on the early gold mines of Los Angeles County, Orange County and Riverside County, which were some of the first gold fields mined in California by early Spanish and Mexican miners before the 49ers came on the scene. Included are not only details on early mining methods in the area, production statistics and geological information, but also the location of the early gold mines that helped make California "The Golden State". 8.5" X 11", 146 ppgs. Retail Price: $12.99

Mines of San Bernadino and Tulare Counties, California - Originally published in 1917, this important publication on California Mining has not been available for nearly a century. This publication includes important information on the early gold mines of San Bernadino and Tulare County, which were some of the first gold fields mined in California by early Spanish and Mexican miners before the 49ers came on the scene. Included are not only details on early mining methods in the area, production statistics and geological information, but also the location of the early gold mines that helped make California "The Golden State". Also included are details on the mining of other minerals such as copper, iron, lead, zinc, manganese, tungsten, vanadium, asbestos, barite, borax, cement, clay, dolomite, fluospar, gem stones, graphite, marble, salines, petroleum, stronium, talc and others. 8.5" X 11", 200 ppgs. Retail Price: $19.99

Chromite Mining in The Klamath Mountains of California and Oregon - Unavailable since 1919, this publication was originally compiled by J.S. Diller of the United States Department of Geological Survey and includes details on the chromite mines of this area of Northern California and Southern Oregon. Also included are insights into the mineralization and other characteristics of this important mining region, as well as the location of historic mines. Also included are insights into chromite mining in Eastern Oregon and Montana. 8.5" X 11", 98 ppgs. Retail Price: $9.99

Mines and Mining in Amador, Calaveras and Tuolumne Counties, California - Unavailable since 1915, this publication was originally compiled by William Tucker and includes details on the mines and mineral resources of this important California mining area. Included are details on the geology, history and important gold mines of the region, as well as insights into other local mineral resources such as asbestos, clay, copper, talc, limestone and others. Also included are insights into the mineralization and other characteristics of this important portion of California's Mother Lode mining region. 8.5" X 11", 198 ppgs. Retail Price: $14.99

The Cerro Gordo Mining District of Inyo County California - Unavailable since 1963, this publication was originally compiled by the United States Department of Interior. Included are insights into the mineralization and other characteristics of this important mining region of Southern California. Topics include the mining of gold and silver in this important mining district in Inyo County, California, including details on the history, production and locations of the Cerro Gordo Mine, the Morning Star Mine, Estelle Tunnel, Charles Lease Tunnel, Ignacio, Hart, Crosscut Tunnel, Sunset, Upper Newtown, Newtown, Ella, Perseverance, Newsboy, Belmont and other silver and gold mines in the Cerro Gordo Mining District. This volume also includes important insights into the fossil record, geologic formations, faults and other aspects of economic geology in this California mining district. **8.5" X 11", 104 ppgs. Retail Price: $10.99**

Mining in Butte, Lassen, Modoc, Sutter and Tehama Counties of California - Unavailable since 1917, this publication was originally compiled by the United States Department of Interior. Included are insights into the mineralization and other characteristics of this important mining region of California. Topics include the mining of asbestos, chromite, gold, diamonds and manganese in Butte County, the mining of gold and copper in the Hayden Hill and Diamond Mountain mining districts of Lassen County, the mining of coal, salt, copper and gold in the High Grade and Winters mining districts of Modoc County, gold mining in Sutter County and the mining of gold, chromite, manganese and copper in Tehama County. This volume also includes the production records and locations of numerous mines in this important mining region. **8.5" X 11", 114 ppgs. Retail Price: $11.99**

Mines of Trinity County California - Originally published in 1965, this important publication on California Mining has not been available for nearly fifty years. This publication includes important information on mines and mining in Trinity County, California, as well insights into the mineralization and geology of this important mining area in Northern California. Included are extensive details on hardrock and placer gold mines and prospects, including charts showing the locations of these historic mines.. **8.5" X 11", 144 ppgs. Retail Price: $12.99**

Mines of Kern County California - Originally published in 1962, this important publication on California Mining has not been available for nearly fifty years. This publication includes important information on mines and mining in Kern County, California, as well insights into the mineralization and geology of this important mining area in California. Included are extensive details on hardrock and placer gold mines and prospects, including charts showing the locations of these historic mines. **8.5" X 11", 398 ppgs. Retail Price: $24.99**

Mines of Calaveras County California - Originally published in 1962, this important publication on California Mining has not been available for nearly fifty years. This publication includes important information on mines and mining in Calaveras County, California, as well insights into the mineralization and geology of this important mining area in Northern California. Included are extensive details on hardrock and placer gold mines and prospects, including charts showing the locations of these historic mines. **8.5" X 11", 236 ppgs. Retail Price: $19.99**

Lode Gold Mining in Grass Valley California - Unavailable since 1940, this publication was originally compiled by the United States Department of Interior. Included are insights into the gold mineralization and other characteristics of this important mining region of Nevada County, California. This volume also includes important insights into the geologic formations, faults and other aspects of economic geology in this California mining district. Of particular interest are the fine details on many hardrock gold mines in the area, including their locations, histories, development and mineralization. Some of the mines featured include the Gold Hill Mine, Massachusetts Hill, Boundary, Peabody, Golden Center, North Star, Omaha, Lone Jack, Homeward Bound, Hartery, Wisconsin, Allison Ranch, Phoenix, Kate Hayes, W.Y.O.D., Empire, Rich Hill, Daisy Hill, Orleans, Sultana, Centennial, Conlin, Ben Franklin, Crown Point and many others. **8.5" X 11", 148 ppgs. Retail Price: $12.99**

Lode Mining in the Alleghany District of Sierra County California - Unavailable since 1913, this publication was originally compiled by the United States Department of Interior. Included are insights into the mineralization and other characteristics of this important mining region of Sierra County. Included are details on the history, production and locations of numerous hardrock gold mines in this famous California area, including the Tightner Mine, Minnie D., Osceola, Eldorado, Twenty One, Sherman, Kenton, Oriental, Rainbow, Plumbago, Irelan, Gold Canyon, North Fork, Federal, Kate Hardy and others. This volume also includes important insights into the fossil record, geologic formations, faults and other aspects of economic geology in this California mining district. **8.5" X 11", 48 ppgs. Retail Price: $7.99**

Six Months In The Gold Mines During The California Gold Rush - Unavailable since 1850, this important work is a first hand account of one "49'ers" personal experience during the great California Gold Rush, shedding important light on one of the most exciting periods in the history of not only California, but also the world. Compiled from journals written between 1847 and 1849 by E. Gould Buffum, a native of New York, "Six Months In The Gold Mines During The California Gold Rush" offers a rare look into the day to day lives of the people who came to California to work in her gold mines when the state was still a great frontier. **8.5" X 11", 290 ppgs. Retail Price: $19.99**

<u>Quartz Mines of the Grass Valley Mining District of California</u> - Unavailable since 1867, this important publication has not been available since those days. This rare publication offers a short dissertation on the early hardrock mines in this important mining district in the California Mother Lode region between the 1850's and 1860's. Also included are hard to find details on the mineralization and locations of these mines, as well as how they were operated in those day. **8.5" X 11"**, **44 ppgs**. Retail Price: $8.99

<u>Gold Rush on the Feather River</u> - First published in 1924, this short publication by G.C. Mansfield sheds important light on the early history of gold mining on the Feather River. Included are rare insights into the first decade of gold mining and the early mining camps of the Feather River during the 1850's. 64 ppgs., 9.99

<u>The Bodie Mining District of California</u> - First published in 1986, it has been unavailable since those days and sheds important light on this famous mining area. Included are the history, characteristics and locations of numerous old mines around the ghost town of Bodie. 64 ppgs, 8.99

<u>Geology and Mineral Resources of the Gasquet Quadrangle of California-Oregon</u> - First published in 1953, it has been unavailable for over a century and sheds important light on the geological features and mineral resources of this portion of Northern California and Southern Oregon. 80 ppgs, 9.99

Alaska Mining Books

<u>Ore Deposits of the Willow Creek Mining District, Alaska</u> - Unavailable since 1954, this hard to find publication includes valuable insights into the Willow Creek Mining District near Hatcher Pass in Alaska. The publication includes insights into the history, geology and locations of the well known mines in the area, including the Gold Cord, Independence, Fern, Mabel, Lonesome, Snowbird, Schroff-O'Neil, High Grade, Marion Twin, Thorpe, Webfoot, Kelly-Willow, Lane, Holland and others. **8.5" X 11"**, **96 ppgs**. Retail Price: $9.99

<u>The Juneau Gold Belt of Alaska</u> - Unavailable since 1906, this hard to find publication includes valuable insights into the gold mines around Juneau, Alaska. The publication includes important details into the history, geology and locations of the well known gold mines and prospects in the area, including those around Windham Bay, Holkham Bay, Port Snettisham, on Grindstone and Rhine Creeks, Gold Creek, Douglas Island, Salmon Creek, Lemon Creek, Nugget Creek, from the Mendenhall River to Berners Bay, McGinnis Creek, Montana Creek, Peterson Creek, Windfall Creek, the Eagle River, Yankee Basin, Yankee Curve, Kowee Creek and elsewhere. Not only are gold placer mines included, but also hardrock gold mines. **8.5" X 11"**, **224 ppgs**. Retail Price: $19.99

<u>Mining in the Jumbo Basin of Alaska</u> - Unavailable since 1953, this hard to find publication includes valuable insights into the mines and geology of the Jumbo Basin. The publication includes important details into the history, geology and locations of the well known gold mines and prospects in the famous Jumbo Basin Mining Region of Alaska. 72 ppgs, 9.99

<u>The Rampart Placer Gold Region of Alaska</u> - Unavailable since 1906, this hard to find publication includes valuable insights into the placer gold mines of the Rampart Mining Region. The publication includes important details into the history, geology and locations of the well known gold mines and prospects in the famous Rampart Mining Region of Alaska. 78 ppgs, 10.99

Arizona Mining Books

<u>Mines and Mining in Northern Yuma County Arizona</u> - Originally published in 1911, this important publication on Arizona Mining has not been available for over a hundred years. Included are rare insights into the gold, silver, copper and quicksilver mines of Yuma County, Arizona together with hard to find maps and photographs. Some of the mines and mining districts featured include the Planet Copper Mine, Mineral Hill, the Clara Consolidated Mine, Viati Mine, Copper Basin prospect, Bowman Mine, Quartz King, Billy Mack, Carnation, the Wardwell and Osbourne, Valensuella Copper, the Mariquita, Colonial Mine, the French American, the New York-Plomosa, Guadalupe, Lead Camp, Mudersbach Copper Camp, Yellow Bird, the Arizona Northern (Salome Strike), Bonanza (Harqua Hala), Golden Eagle, Hercules, Socorro and others. **8.5" X 11"**, **144 ppgs**. Retail Price: $11.99

<u>The Aravaipa and Stanley Mining Districts of Graham County Arizona</u> - Originally published in 1925, this important publication on Arizona Mining has not been available for nearly ninety years. Included are rare insights into the gold and silver mines of these two important mining districts, together with hard to find maps. **8.5" X 11"**, **140 ppgs**. Retail Price: $11.99

Gold in the Gold Basin and Lost Basin Mining Districts of Mohave County, Arizona - This volume contains rare insights into the geology and gold mineralization of the Gold Basin and Lost Basin Mining Districts of Mohave County, Arizona that will be of benefit to miners and prospectors. Also included is a significant body of information on the gold mines and prospects of this portion of Arizona. This volume is lavishly illustrated with rare photos and mining maps. **8.5" X 11", 188 ppgs. Retail Price: $19.99**

Mines of the Jerome and Bradshaw Mountains of Arizona - This important publication on Arizona Mining has not been available for ninety years. This volume contains rare insights into the geology and ore deposits of the Jerome and Bradshaw Mountains of Arizona that will be of benefit to miners and prospectors who work those areas. Included is a significant body of information on the mines and prospects of the Verde, Black Hills, Cherry Creek, Prescott, Walker, Groom Creek, Hassayampa, Bigbug, Turkey Creek, Agua Fria, Black Canyon, Peck, Tiger, Pine Grove, Bradshaw, Tintop, Humbug and Castle Creek Mining Districts. This volume is lavishly illustrated with rare photos and mining maps. **8.5" X 11", 218 ppgs. Retail Price: $19.99**

The Ajo Mining District of Pima County Arizona - This important publication on Arizona Mining has not been available for nearly seventy years. This volume contains rare insights into the geology and mineralization of the Ajo Mining District in Pima County, Arizona and in particular the famous New Cornelia Mine. **8.5" X 11", 126 ppgs. Retail Price: $11.99**

Mining in the Santa Rita and Patagonia Mountains of Arizona - Originally published in 1915, this important publication on Arizona Mining has not been available for nearly a century. Included are rare insights into hundreds of gold, silver, copper and other mines in this famous Arizona mining area. Details include the locations, geology, history, production and other facts of the mines of this region. **8.5" X 11", 394 ppgs. Retail Price: $24.99**

Mining in the Bisbee Quadrangle of Arizona - Originally published in 1906, this important publication on Arizona Mining has not been available for nearly a century. Included are rare insights into hundreds of gold, silver, copper and other mines in this famous Arizona mining area. Details include the locations, geology, history, production and other facts of the mines of this important mining region. **8.5" X 11", 188 ppgs. Retail Price: $14.99**

Placer Gold Mining in Arizona - Unavailable since 1922, this hard to find publication includes valuable insights into the placer gold mines of the Arizona. Originally released as "Placer Gold of Arizona", despite its small size, this publication includes important details into the history, geology and locations of the well known placer gold mines and prospects in the State of Arizona. 48 ppgs, 8.99

Gold and Copper Mining near Payson, Arizona - Written in 1915, this hard to find publication includes valuable insights into the gold and copper mining industry of Arizona. Highlighted here are the gold and copper mines near Payson, Arizona. 68 ppgs, 8.99

Lode Gold Mining in Arizona - Unavailable since 1934, this hard to find publication, originally released as "Arizona Lode Gold Mines and Gold Mining" includes valuable insights into the gold mining industry of Arizona. Included are valuable insights into over 150 hardrock gold mines in over 30 different mining districts in Arizona. 278 ppgs, 21.99

Mining in the Dragoon Quadrangle of Cochise County, Arizona - Unavailable since 1964, this hard to find publication includes valuable insights into the mines of the Dragoon Quadrangle Mining Region. The publication includes important details into the history, geology and locations of the well known mines and prospects in this famous mining region of Arizona. 224 ppgs., 19.99

Directory of Operating Mines in Arizona in 1915 - Unavailable since 1916, this hard to find publication includes valuable insights into the mines of Arizona. This small publication includes a complete list of the mines that were operating in the State of Arizona during 1915 and includes details such as general location, owners and some basic facts about each mining operation.52 ppgs. 8.99

Arizona Ore Deposits - Unavailable since 1938, this hard to find publication includes valuable insights into some ore deposits of Arizona. Included are valuable insights into the formation and characteristics of valuable ore deposits in the Jerome, Miami, Inspiration, Clifton, Morenci, Ray, Ajo, Eureka, Tombstone and Magma mining districts. Included are details into some of the major gold, silver and copper mines of these important Arizona mining areas. 160 ppgs, 14.99

Montana Mining Books

A History of Butte Montana: The World's Greatest Mining Camp - First published in 1900 by H.C. Freeman, this important publication sheds a bright light on one of the most important mining areas in the history of The West. Together with his insights, as well as rare photographs of the periods, Harry Freeman describes Butte and its vicinity from its early beginnings, right up to its flush years when copper flowed from its mines like a river. At the time of publication, Butte, Montana was known worldwide as "The Richest Mining Spot On Earth" and produced not only vast amounts of copper, but also silver, gold and other metals from its mines. Freeman illustrates, with great detail, the most important mines in the vicinity of Butte, providing rare details on their owners, their history and most importantly, how the mines operated and how their treasures were extracted. Of particular interest are the dozens of rare photographs that depict mines such as the famous Anaconda, the Silver Bow, the Smoke House, Moose, Paulin, Buffalo, Little Minah, the Mountain Consolidated, West Greyrock, Cora, the Green Mountain, Diamond, Bell, Parnell, the Neversweat, Nipper, Original and many others. 8.5" X 11", 142 ppgs. Retail Price: $12.99

The Butte Mining District of Montana - This important publication on Montana Mining has not been available for over a century. Included are rare insights into the gold, copper and silver mines of Butte, Montana together with hard to find maps and photographs. Some of the topics include the early history of gold, silver and copper mining in the Butte area, insight into the geology of its mining areas, the local distribution of gold, silver and copper ores, as well their composition and how to identify them. Also included are detailed facts about the mines in the Butte Mining District, including the famous Anaconda Mine, Gagnon, Parrot, Blue Vein, Moscow, Poulin, Stella, Buffalo, Green Mountain, Wake Up Jim, the Diamond-Bell Group, Mountain Consolidated, East Greyrock, West Greyrock, Snowball, Corra, Speculator, Adirondack, Miners Union, the Jessie-Edith May Group, Otisco, Iduna, Colorado, Lizzie, Cambers, Anderson, Hesperus, Preferencia and dozens of others. 8.5" X 11", 298 ppgs. Retail Price: $24.99

Mines of the Helena Mining Region of Montana - This important publication on Montana Mining has not been available for over a century. Included are rare insights into the gold, copper and silver mines of the vicinity of Helena, Montana, including the Marysville Mining District, Elliston Mining District, Rimini Mining District, Helena Mining District, Clancy Mining District, Wickes Mining District, Boulder and Basin Mining Districts and the Elkhorn Mining District. Some of the topics include the early history of gold, silver and copper mining in the Helena area, insight into the geology of its mining areas, the local distribution of gold, silver and copper ores, as well their composition and how to identify them. Also included are detailed facts, history, geology and locations of over one hundred gold, silver and copper mines in the area . 8.5" X 11", 162 ppgs, Retail Price: $14.99

Mines and Geology of the Garnet Range of Montana - This important publication on Montana Mining has not been available for over a century. Included are rare insights into the gold, copper and silver mines of the vicinity of this important mining area of Montana. Some of the topics include the early history of gold, silver and copper mining in the Garnet Mountains, insight into the geology of its mining areas, the local distribution of gold, silver and copper ores, as well their composition and how to identify them. Also included are detailed facts, history, geology and locations of numerous gold, silver and copper mines in the area . 8.5" X 11", 100 ppgs, Retail Price: $11.99

Mines and Geology of the Philipsburg Quadrangle of Montana - This important publication on Montana Mining has not been available for over a century. Included are rare insights into the gold, copper and silver mines of the vicinity of this important mining area of Montana. Some of the topics include the early history of gold, silver and copper mining in the Philipsburg Quadrangle, insight into the geology of its mining areas, the local distribution of gold, silver and copper ores, as well their composition and how to identify them. Also included are detailed facts, history, geology and locations of over one hundred gold, silver and copper mines in the area 8.5" X 11", 290 ppgs, Retail Price: $24.99

Geology of the Marysville Mining District of Montana - Included are rare insights into the mining geology of the Marysville Mining District. Some of the topics include the early history of gold, silver and copper mining in the area, insight into the geology of its mining areas, the local distribution of gold, silver and copper ores, as well their composition and how to identify them. Also included are detailed facts, history, geology and locations of gold, silver and copper mines in the area 8.5" X 11", 198 ppgs, Retail Price: $19.99

The Geology and Mines of Northern Idaho and North Western Montana- See listing under Idaho.

The History of Gold Dredging in Montana - Unavailable since 1916, this important publication was originally published by the Us Bureau of Mines and has been unavailable for a century. A century and more ago, giant dredging machines dug in Montana's rivers and creeks in search of illusive golden riches. First appearing in California in the 1850's, gold dredges finally reached their peak of development in Siberia and New Zealand before becoming popular again in the United States. This book offers a unique historical perspective on the gold dredges that once operated in Montana. This book on Montana mining history is lavishly illustrated with dozens of rare historic photos gold dredges that once operated in Montana, as well as hard to locate plans on how these dredges were designed. 120 ppgs., 11.99

Nevada Mining Books

The Bull Frog Mining District of Nevada - Unavailable since 1910, this publication was originally compiled by the United States Department of Interior. This volume also includes important insights into the geologic formations, faults and other aspects of economic geology in this Nevada mining district. Of particular interest are the fine details on many mines in the area, including their locations, histories, development and mineralization. Some of the mines featured include the National Bank Mine, Providence, Gibraltor, Tramps, Denver, Original Bullfrog, Gold Bar, Mayflower, Homestake-King and other mines and prospects. **8.5" X 11", 152 ppgs, Retail Price: $14.99**

History of the Comstock Lode - Unavailable since 1876, this publication was originally released by John Wiley & Sons. This volume also includes important insights into the famous Comstock Lode of Nevada that represented the first major silver discovery in the United States. During its spectacular run, the Comstock produced over 192 million ounces of silver and 8.2 million ounces of gold. Not only did the Comstock result in one of the largest mining rushes in history and yield immense fortunes for its owners, but it made important contributions to the development of the State of Nevada, as well as neighboring California. Included here are important details on not only the early development and history of the Comstock, but also rare early insight into its mines, ore and its geology.8.5" X 11", 244 ppgs, Retail Price: $19.99

The Pioche Mining District of Nevada - First published in 1932, it has been unavailable for over a century and sheds important light on the mining history of Nevada. Some of the topics include the history of mining in this district, as well as the characteristics of its mineral and ore deposits. Also included are insights into the history, production, characteristics and locations of numerous mines in the area. Some of the mines include the Combined Metals, Pioche, Ely Valley, No. 10, Poorman, Wide Awake, Alps, Prince, Virginia Louise, Half Moon, Abe Lincoln, Fairview, Bristol Silver, National, Vesuvius, Inman, Tempest, Hillside, Jackrabbit, Lucky Star, Fortuna, Mendha, Manhattan, Hamburg, Comet, Lyndon and others. 108 ppgs 10.99

The Yerington Mining District of Nevada - First published in 1932, it has been unavailable for over a century and sheds important light on the mining history of Nevada. Some of the topics include the history of mining in this district, as well as the characteristics of its mineral and ore deposits. Also included are insights into the history, production, characteristics and locations of numerous mines in the area. Some of the mines include the Bluestone, Mason Valley, Malachite, McConnell, Greenwood, Western Nevada, Ludwig, Douglas Hill, Casting Copper, Montana-Yerington, Empire, Jim Beatty, Terry and McFarland, Blue Jay and others. 92 ppgs, 10.99

The Genesis of the Ores of Tonopah Nevada - Unavailable since 1918, this hard to find publication includes valuable insights into the gold mines around Tonopah, Nevada. The publication includes important details into the geology of mines in the Tonopah Mining District of Nevada. 90 ppgs, 10.99

Mining Camps of Elko, Lander and Eureka Counties Nevada - Unavailable since 1910, this hard to find publication includes valuable insights into the mining camps of Elko, Lander and Eureka Counties, Nevada. The publication includes important details into the history of mines and mining in these three Nevada counties. 154 ppgs, 12.99

Ore Deposits of the Bullfrog Quadrangle - Unavailable since 1964 and released as "Geology of Bullfrog Quadrangle and Ore Deposits Related to Bullfrog Hills Caldera, Nye County, Nevada and Inyo County, California". The publication includes important details into the geology of mines in the Bullfrog Quadrangle of Nye County, Nevada and Inyo County, California. 52 ppgs, 9.99

Mining in Eureka County Nevada - Unavailable since 1879, this hard to find publication includes valuable insights into the early mining history off Eureka County, Nevada. The publication includes important details into the early history of the mines of Eureka County, as well as their development, production and how their ores were treated. Also included are details on the 1872 Mining Act, as well as the local rules, regulations and customs of the miners in Eureka County.134 ppgs, 12.99

Colorado Mining Books

Ores of The Leadville Mining District - Unavailable since 1926, this publication was originally compiled by the United States Department of Interior. This volume also includes important insights into the ores and mineralization of the Leadville Mining District in Colorado. Topics include historic ore prospecting methods, local geology, insights into ore veins and stockworks, the local trend and distribution of ore channels, reverse faults, shattered rock above replacement ore bodies, mineral enrichment in oxidized and sulphide zones and more. **8.5" X 11", 66 ppgs, Retail Price: $8.99**

Mining in Colorado - Unavailable since 1926, this publication was originally compiled by the United States Department of Interior. This volume also includes important insights into the mining history of Colorado from its early beginnings in the 1850's right up to the mid 1920's. Not only is Colorado's gold mining heritage included, but also its silver, copper, lead and zinc mining industry. Each mining area is treated separately, detailing the development of Colorado's mines on a county by county basis. **8.5" X 11", 284 ppgs, Retail Price: $19.99**

Gold Mining in Gilpin County Colorado - Unavailable since 1876, this publication was originally compiled by the Register Steam Printing House of Central City, Colorado. A rare glimpse at the gold mining history and early mines of Gilpin County, Colorado from their first discovery in the 1850's up to the "flush years" of the mid 1870's. Of particular interest is the history of the discovery of gold in Gilpin County and details about the men who made those first strikes. Special focus is given to the early gold mines and first mining districts of the area, many of which are not detailed in other books on Colorado's gold mining history. **8.5" X 11", 156 ppgs, Retail Price: $12.99**

Mining in the Gold Brick Mining District of Colorado - Important insights into the history of the Gold Brick Mining District, as well as its local geography and economic geology. Also included are the histories and locations of historic mines in this important Colorado Mining District, including the Cortland, Carter, Raymond, Gold Links, Sacramento, Bassick, Sandy Hook, Chronicle, Grand Prize, Chloride, Granite Mountain, Lucille, Gray Mountain, Hilltop, Maggie Mitchell, Silver Islet, Revenue, Roosevelt, Carbonate King and others. In addition to hardrock mining, are also included are details on gold placer mining in this portion of Colorado. **8.5" X 11", 140 ppgs, Retail Price: $12.99**

Ore Deposits of the London Fault of Colorado - First published in 1941, it has been unavailable since those days and sheds important light on the mines and mineral deposits of the London Fault in Central Colorado's Alma Mining District. This publication sheds important light on the gold veins and lead-silver deposits of the Alma Mining District. Included are geologic details on the London Mine, American Mine, Havigorst Tunnel, Ophir Mine, Mosher Tunnel, London-Butte Mine, Venture Shaft, Hard-To-Beat Mine, Oliver Twist Tunnel, Sacramento Mine, Mudsill Mine, Sherwood Mine, Wagner, Barcoe Tunnel and other mines in this important mining region. 110 ppgs., 10.99

The Mines of Colorado - First published in 1867, it has been unavailable since those days and sheds important light on Colorado's early mining history. Written shortly after the events took place, this publication sheds important light on the Pike's Peak Gold Rush, the discovery of gold on Ralston Creek and Dry Creek in the 1850's, as well as details on the first wave of miners into Colorado and their trials and tribulations as they crossed the Great Plains. Also included are details on early discoveries of lode gold in the mountainous regions of Colorado, details on the early mines hardrock and placer mines, and much more. It is a veritable treasure trove on Colorado's early mining history and will be of great importance to anyone who is interested in the mining of gold or other minerals in Colorado, as well as those interested in the history of the state. 478 ppgs., 29.99

The La Plata Mining District of Colorado - Originally titled "Geology and Ore Deposits in the Vicinity of the La Plata District of Colorado" and first published in 1949, it has been unavailable since those days and sheds important light on the mines and mineral deposits of the La Plata Mining District of Colorado. 214 ppgs., 19.99

Washington Mining Books

The Republic Mining District of Washington - Unavailable since 1910, this important publication was originally published by the Washington Geologic Survey and has been unavailable for a century. Topics include the geology, rock formations and the formation of ore deposits in this important mining area of Washington State. Also included are hard to find details on the geology, history and locations of dozens of mines in the area. Some of the mines featured include the New Republic Mine, Ben Hur, Morning Glory, the South Republic Mine, Quilp, Surprise, Black Tail, Lone Pine, San Poil, Mountain Lion, Tom Thumb, Elcaliph and many others. 8.5″ X 11″, 94 ppgs, **Retail Price: $10.99**

The Myers Creek and Nighthawk Mining Districts of Washington - Unavailable since 1911, this important publication was originally published by the Washington Geologic Survey and has been unavailable for a century. Topics include the geology, rock formations and the formation of ore deposits in these important mining areas of Washington State. Also included are hard to find details on the geology, history and locations of dozens of mines in the area. Some of the mines featured include the Grant Mine, Monterey, Nip and Tuck, Myers Creek, Number Nine, Neutral, Rainbow, Aztec, Crystal Butte, Apex, Butcher Boy, Molson, Mad River, Olentangy, Delate, Kelsey, Golden Chariot, Okanogan, Ohio, Forty-Ninth Parallel, Nighthawk, Favorite, Little Chopaka, Summit, Number One, California, Peerless, Caaba, Prize Group, Ruby, Mountain Sheep, Golden Zone, Rich Bar, Similkameen, Kimberly, Triune, Hiawatha, Trinity, Hornsilver, Maquae, Bellevue, Bullfrog, Palmer Lake, Ivanhoe, Copper World and many others. 8.5″ X 11″, 136 ppgs, **Retail Price: $12.99**

The Blewett Mining District of Washington - Unavailable since 1911, this important publication was originally published by the Washington Geologic Survey and has been unavailable for a century. Topics include the geology, rock formations and the formation of ore deposits in this important mining area of Washington State. Also included are hard to find details on the geology, history and locations of dozens of mines in the area. Some of the mines featured include the Washington Meteor, Alta Vista, Pole Pick, Blinn, North Star, Golden Eagle, Tip Top, Wilder, Golden Guinea, Lucky Queen, Blue Bell, Prospect, Homestake, Lone Rock, Johnson, and others. 8.5″ X 11″, 134 ppgs, **Retail Price: $12.99**

Silver Mining In Washington - Unavailable since 1955, this important publication was originally published by the Washington Geologic Survey. Featured are the hard to find locations and details pertaining to Washington's silver mines. 8.5″ X 11″, 180 ppgs, **Retail Price: $15.99**

The Mines of Snohomish County Washington - Unavailable since 1942, this important publication was originally published by the Washington Geologic Survey and has been unavailable for seventy years. Featured are details on a large number of gold, silver, copper, lead and other metallic mineral mines. Included are the locations of each historic mine, along with information on the commodity produced. 8.5″ X 11″, 98 ppgs, **Retail Price: $10.99**

The Mines of Chelan County Washington - Unavailable since 1943, this important publication was originally published by the Washington Geologic Survey and has been unavailable for seventy years. Featured are details on a large number of gold, silver, copper, lead and other metallic mineral mines. Included are the locations of each historic mine, along with information on the commodity. 8.5″ X 11″, 88 ppgs, **Retail Price: $9.99**

Metal Mines of Washington - Unavailable since 1921, this important publication was originally published by the Washington Geologic Survey and has been unavailable for nearly ninety years. Widely considered a masterpiece on the Washington Mining Industry, "Metal Mines of Washington" sheds light on the important details of Washington's early mining years. Featured are details on hundreds of gold, silver, copper, lead and other metallic mineral mines. Included are hard to find details on the mineral resources of this state, as well as the locations of historic mines. Lavishly illustrated with maps and historic photos and complete with a glossary to explain any technical terms found in the text, this is one of the most important works on mining in the State of Washington. No prospector or miner should be without it if they are interested in mining in Washington. 8.5″ X 11″, 396 ppgs, **Retail Price: $24.99**

Gem Stones In Washington - Unavailable since 1949, this important publication was originally published by the Washington Geologic Survey and has been unavailable since first published. Included are details on where to find naturally occurring gem stones in the State of Washington, including quartz crystal, amethyst, smoky quartz, milky quartz, agates, bloodstone, carnelian, chert, flint, jasper, onyx, petrified wood, opal, fire opal, hyalite and others. 8.5″ X 11″, 54 ppgs, **Retail Price: $8.99**

The Covada Mining District of Washington - Unavailable since 1913, this important publication was originally published by the Washington Geologic Survey and has been unavailable for a century. Topics include the geology, rock formations and the formation of ore deposits in this important mining area of Washington State. Also included are hard to find details on the geology, history and locations of dozens of mines in the area. Some of the mines featured include the Admiral, Advance, Algonkian, Big Bug, Big Chief, Big Joker, Black Hawk, Black Tail, Black Thorn, Captain, Cherokee Strip, Colorado, Dan Patch, Dead Shot, Etta, Good Ore, Greasy Run, Great Scott, Idora, IXL, Jay Bird, Kentucky Bell, King Solomon, Laurel, Laura S, Little Jay, Meteor, Neglected, Northern Light, Old Nell, Plymouth Rock, Polaris, Quandary, Reserve, Shoo Fly, Silver Plume, Three Pines, Vernie, White Rose and dozens of others. 8.5″ X 11″, 114 ppgs, **Retail Price: $10.99**

The Index Mining District of Washington - Unavailable since 1912, this important publication was originally published by the Washington Geologic Survey and has been unavailable for a century. Topics include the geology, rock formations and the formation of ore deposits in this important mining area of Washington State. Also included are hard to find details on the geology, history and locations of dozens of mines in the area. Some of the mines featured include the Sunset, Non-Pareil, Ethel Consolidated, Kittaning, Merchant, Homestead, Co-operative, Lost Creek, Uncle Sam, Calumet, Florence-Rae, Bitter Creek, Index Peacock, Gunn Peak, Helena, North Star, Buckeye. Copper Bell, Red Cross and others. **8.5" X 11", 114 ppgs, Retail Price: $11.99**

Mining & Mineral Resources of Stevens County Washington - Unavailable since 1920, this important publication was originally published by the Washington Geologic Survey and has been unavailable for a century. Topics include the geology, rock formations and the formation of ore deposits in these important mining areas of Washington State. Also included are hard to find details on the geology, history and locations of hundreds of mines in the area. **8.5" X 11", 372 ppgs, Retail Price: $24.99**

The Mines and Geology of the Loomis Quadrangle Okanogan County, Washington - Unavailable since 1972, this important publication was originally published by the Washington Geologic Survey and has been unavailable for a century. Topics include the geology, rock formations and the formation of ore deposits in this important mining area of Washington State. Also included are hard to find details on the geology, history and locations of dozens of gold, copper, silver and other mines in the area. **8.5" X 11", 150 ppgs, Retail Price: $12.99**

The Conconully Mining District of Okanogan County Washington - Unavailable since 1973, this important publication was originally published by the Washington Geologic Survey and has been unavailable for a century. Topics include the geology, rock formations and the formation of ore deposits in this important mining area of Washington State, which also includes Salmon Creek, Blue Lake and Galena. Also included are hard to find details on the geology, mining history and locations of dozens of mines in the area. Some of the mines include Arlington, Fourth of July, Sonny Boy, First Thought, Last Chance, War Eagle-Peacock, Wheeler, Mohawk, Lone Star, Woo Loo Moo Loo, Keystone, Hughes, Plant-Callahan, Johnny Boy, Leuena, Gubser, John Arthur, Tough Nut, Homestake, Key and many others **8.5" X 11", 68 ppgs, Retail Price: $8.99**

Wyoming Mining Books

Mining in the Laramie Basin of Wyoming - Unavailable since 1909, this publication was originally compiled by the United States Department of Interior. Also included are insights into the mineralization and other characteristics of this important mining region, especially in regards to coal, limestone, gypsum, bentonite clay, cement, sand, clay and copper. **8.5" X 11", 104 ppgs, Retail Price: $11.99**

New Mexico Mining Books

The Mogollon Mining District of New Mexico - Unavailable since 1927, this important publication was originally published by the US Department of Interior and has been unavailable for 80 years. Topics include the geology, rock formations and the formation of ore deposits in this important mining area in New Mexico. Of particular focus is information on the history and production of the ore deposits in this area, their form and structure, vein filling, their paragenesis, origins and ore shoots, as well as oxidation and supergene enrichment. Also included are hard to find details, including the descriptions and locations of numerous gold, silver and other types of mines, including the Eureka, Pacific, South Alpine, Great Western, Enterprise, Buffalo, Mountain View, Floride, Gold Dust, Last Chance, Deadwood, Confidence, Maud S., Deep Down, Little Fanney, Trilby, Johnson, Alberta, Comet, Golden Eagle, Cooney, Queen, the Iron Crown, Eberle, Clifton, Andrew Jackson mine, Mascot and others. **8.5" X 11", 144 ppgs, Retail Price: $12.99**

The Percha Mining District of Kingston New Mexico - Unavailable since 1883, this important publication was originally published by the Kingston Tribune and has been unavailable for over one hundred and thirty five years. Having been written during the earliest years of gold and silver mining in the Percha Mining District, unlike other books on the subject, this work offers the unique perspective of having actually been written while the early mining history of this area was still being made. In fact, the work was written so early in the development of this area that many of the notable mines in the Percha District were less than a few years old and were still being operated by their original discoverers with the same enthusiasm as when they were first located. Included are hard to find details on the very earliest gold and silver mines of this important mining district near Kingston in Sierra County, New Mexico. **8.5" X 11", 68 ppgs, Retail Price: $9.99**

East Coast Mining Books

The Gold Fields of the Southern Appalachians - Unavailable since 1895, this important publication was originally published by the US Department of Interior and has been unavailable for nearly 120 years. Topics include the geology, rock formations and the formation of ore deposits in this important mining area of the American South. Of particular focus is information on the history and statistics of the ore deposits in this area, their form and structure and veins. Also included are details on the placer gold deposits of the region. The gold fields of the Georgian Belt, Carolinian Belt and the South Mountain Mining District of North Carolina are all treated in descriptive detail. Included are hard to find details, including the descriptions and locations of numerous gold mines in Georgia, North Carolina and elsewhere in the American South. Also included are details on the gold belts of the British Maritime Provinces and the Green Mountains. **8.5" X 11", 104 ppgs, Retail Price: $9.99**

Gold Rush Tales Series

Millions in Siskiyou County Gold - In this first volume of the "Gold Rush Tales" series, leading mining historian and editor Kerby Jackson, introduces us to the story of how millions of dollars worth of gold was discovered in Siskiyou County during the California Gold Rush. Lavishly illustrated with photos from the 19th Century, this hard to find information was first published in 1897 and sheds important light onto the gold rush era in Siskiyou County, California and the experiences of the men who dug for the gold and actually found it. **8.5" X 11", 82 ppgs, Retail Price: $9.99**

The California Rand in the Days of '49 - In this second volume of the "Gold Rush Tales" series, leading mining historian and editor Kerby Jackson, introduces us to four tales from the California Gold Rush. Lavishly illustrated with photos from the 19th Century, this hard to find information was first published in 1890's and includes the stories of "California's Rand", details about Chinese miners, how one early miner named Baker struck it rich and also the story of Alphonzo Bowers, who invented the first hydraulic gold dredge. **8.5" X 11", 54 ppgs, Retail Price: $9.99**

More Mining Books

Prospecting and Developing A Small Mine - Topics covered include the classification of varying ores, how to take a proper ore sample, the proper reduction of ore samples, alluvial sampling, how to understand geology as it is applied to prospecting and mining, prospecting procedures, methods of ore treatment, the application of drilling and blasting in a small mine and other topics that the small scale miner will find of benefit. **8.5" X 11", 112 ppgs, Retail Price: $11.99**

Timbering For Small Underground Mines - Topics covered include the selection of caps and posts, the treatment of mine timbers, how to install mine timbers, repairing damaged timbers, use of drift supports, headboards, squeeze sets, ore chute construction, mine cribbing, square set timbering methods, the use of steel and concrete sets and other topics that the small underground miner will find of benefit. This volume also includes twenty eight illustrations depicting the proper construction of mine timbering and support systems that greatly enhance the practical usability of the information contained in this small book. **8.5" X 11", 88 ppgs. Retail Price: $10.99**

Timbering and Mining - A classic mining publication on Hard Rock Mining by W.H. Storms. Unavailable since 1909, this rare publication provides an in depth look at American methods of underground mine timbering and mining methods. Topics include the selection and preservation of mine timbers, drifting and drift sets, driving in running ground, structural steel in mine workings, timbering drifts in gravel mines, timbering methods for driving shafts, positioning drill holes in shafts, timbering stations at shafts, drainage, mining large ore bodies by means of open cuts or by the "Glory Hole" system, stoping out ore in flat or low lying veins, use of the "Caving System", stoping in swelling ground, how to stope out large ore bodies, Square Set timbering on the Comstock and its modifications by California miners, the construction of ore chutes, stoping ore bodies by use of the "Block System", how to work dangerous ground, information on the "Delprat System" of stoping without mine timbers, construction and use of headframes and much more. This volume provides a reference into not only practical methods of mining and timbering that may be employed in narrow vein mining by small miners today, but also rare insights into how mines were being worked at the turn of the 19th Century. **8.5" X 11", 288 ppgs. Retail Price: $24.99**

A Study of Ore Deposits For The Practical Miner - Mining historian Kerby Jackson introduces us to a classic mining publication on ore deposits by J.P. Wallace. First published in 1908, it has been unavailable for over a century. Included are important insights into the properties of minerals and their identification, on the occurrence and origin of gold, on gold alloys, insights into gold bearing sulfides such as pyrites and arsenopyrites, on gold bearing vanadium, gold and silver tellurides, lead and mercury tellurides, on silver ores, platinum and iridium, mercury ores, copper ores, lead ores, zinc ores, iron ores, chromium ores, manganese ores, nickel ores, tin ores, tungsten ores and others. Also included are facts regarding rock forming minerals, their composition and occurrences, on igneous, sedimentary, metamorphic and intrusive rocks, as well as how they are geologically disturbed by dikes, flows and faults, as well as the effects of these geologic actions and why they are important to the miner. Written specifically with the common miner and prospector in mind, the book will help to unlock the earth's hidden wealth for you and is written in a simple and concise language that anyone can understand. **8.5" X 11", 366 ppgs. Retail Price: $24.99**

Mine Drainage - Unavailable since 1896, this rare publication provides an in depth look at American methods of underground mine drainage and mining pump systems. This volume provides a reference into not only practical methods of mining drainage that may be employed in narrow vein mining by small miners today, but also rare insights into how mines were being worked at the turn of the 19th Century. **8.5" X 11", 218 ppgs. Retail Price: $24.99**

Fire Assaying Gold, Silver and Lead Ores - Unavailable since 1907, this important publication was originally published by the Mining and Scientific Press and was designed to introduce miners and prospectors of gold, silver and lead to the art of fire assaying. Topics include the fire assaying of ores and products containing gold, silver and lead; the sampling and preparation of ore for an assay; care of the assay office, assay furnaces; crucibles and scorifiers; assay balances; metallic ores; scorification assays; cupelling; parting' crucible assays, the roasting of ores and more. This classic provides a time honored method of assaying put forward in a clear, concise and easy to understand language that will make it a benefit to even beginners. **8.5" X 11", 96 ppgs. Retail Price: $11.99**

Methods of Mine Timbering - Originally published in 1896, this important publication on mining engineering has not been available for nearly a century. Included are rare insights into historical methods of timbering structural support that were used in underground metal mines during the California that still have a practical application for the small scale hardrock miner of today. **8.5" X 11", 94 ppgs. Retail Price: $10.99**

The Enrichment of Copper Sulfide Ores - First published in 1913, it has been unavailable for over a century. Topics include the definition and types of ore enrichment, the oxidation of copper ores, the precipitation of metallic sulfides. Also included are the results of dozens of lab experiments pertaining to the enrichment of sulfide ores that will be of interest to the practical hard rock mine operator in his efforts to release the metallic bounty from his mine's ore. **8.5" X 11", 92 ppgs. Retail Price: $9.99**

A Study of Magmatic Sulfide Ores - Unavailable since 1914, this rare publication provides an in depth look at magmatic sulfide ores. Some of the topics included are the definition and classification of magmatic ores, descriptions of some magmatic sulfide ore deposits known at the time of publication including copper and nickel bearing pyrrohitic ore bodies, chalcopyrite-bornite deposits, pyritic deposits, magnetite-ileminite deposits, chromite deposits and magmatic iron ore deposits. Also included are details on how to recognize these types of ore deposits while prospecting for valuable hardrock minerals. **8.5" X 11", 138 ppgs. Retail Price: $11.99**

The Cyanide Process of Gold Recovery - Unavailable since 1894 and released under the name "The Cyanide Process: Its Practical Application and Economical Results", this rare publication provides an in depth look at the early use of cyanide leaching for gold recovery from hardrock mine ores. This volume provides a reference into the early development and use of cyanide leaching to recover gold. **8.5" X 11", 162 ppgs. Retail Price: $14.99**

California Gold Milling Practices - Unavailable since 1895 and released under the name "California Gold Practices", this rare publication provides an in depth look at early methods of milling used to reduce gold ores in California during the late 19th century. This volume provides a reference into the early development and use of milling equipment during the earliest years of the California Gold Rush up to the age of the Industrial Revolution. Much of the information still applies today and will be of use to small scale miners engaging in hardrock mining. **8.5" X 11", 104 ppgs. Retail Price: $10.99**

Leaching Gold and Silver Ores With The Plattner and Kiss Processes - Mining historian Kerby Jackson introduces us to a classic mining publication on the evaluation and examination of mines and prospects by C.H. Aaron. First published in 1881, it has been unavailable for over a century and sheds important light on the leaching of gold and silver ores with the Plattner and Kiss processes. **8.5" X 11", 204 ppgs. Retail Price: $15.99**

The Metallurgy of Lead and the Desilverization of Base Bullion - First published in 1896, it has been unavailable for over a century and sheds important light on the the recovery of silver from lead based ores. Some of the topics include the properties of lead and some of its compounds, lead ores such as galenite, anglesite, cerussite and others, the distribution of lead ores throughout the United States and the sampling and assaying of lead ores. Also covered is the metallurgical treatment of lead ores, as well as the desilverization of lead by the Pattinson Process and the Parkes Process. Hofman's text has long been considered one of the most important early works on the recovery of silver from lead based ores. 8.5" X 11", 452 ppgs. Retail Price: $29.99

Ore Sampling For Small Scale Miners - First published in 1916, it has been unavailable for over a century and sheds important light on historic methods of ore sampling in hardrock mines. Topics include how to take correct ore samples and the conditions that affect sampling, such as their subdivision and uniformity. Particular detail is given to methods of hand sampling ore bodies by grab sample, pipe sample and coning, as well as sampling by mechanical methods. Also given are insights into the screening, drying and grinding processes to achieve the most consistent sample results and much more. 8.5" X 11", 124 ppgs. Retail Price: $12.99

The Extraction of Silver, Copper and Tin from Ores - First published in 1896, it has been unavailable for over a century and sheds important light on how historic miners recovered silver, copper and tin from their mining operations. The book is split into three sections, including a discussion on the Lixiviation of Silver Ores, the mining and treatment of copper ores as practiced at Tharsis, Spain and the smelting of tin as it was practiced by metallurgists at Pulo Brani, Singapore. Also included is an overview and analysis of these historic metal recovery methods that will be of benefit to those interested in the extraction of silver, copper and tin from small mines. 8.5" X 11", 118 ppgs. Retail Price: $14.99

The Roasting of Gold and Silver Ores - First published in 1880, it has been unavailable for over a century and sheds important light on how historic miners recovered gold and silver rom their mining operations. Topics include details on the most important silver and free milling gold ores, methods of desulphurization of ores, methods of deoxidation, the chlorination of ores, methods and details on roasting gold and silver ores, notes on furnaces and more. Also included are details on numerous methods of gold and silver recovery, including the Ottokar Hofman's Process, the Patera Process, Kiss Process, Augustin Process, Ziervogel Process and others. 8.5" X 11", 178 ppgs. Retail Price: $19.99

The Examination of Mines and Prospects - First published in 1912, it has been unavailable for over a century and sheds important light on how to examine and evaluate hardrock mines, prospects and lode mining claims. Sections include Mining Examinations, Structural Geology, Structural Features of Ore Deposits, Primary Ores and their Distribution, Types of Primary Ore Deposits, Primary Ore Shoots, The Primary Alteration of Wall Rocks, Alterations by Surface Agencies, Residual Ores and their Distribution, Secondary Ores and Ore Shoots and Vein Outcrops. This hard to find information is a must for those who are interested in owning a mine or who already own a lode mining claim and wish to succeed at quartz mining. 8.5" X 11", 250 ppgs. Retail Price: $19.99

Garnets: Their Mining, Milling and Utilization - First published in 1925, it has been unavailable since those days and sheds important light on the mining, milling and utilization of garnets. Included are details on the characteristics of garnets, where they are found and how they were mined. 78 ppgs, 10.99

Gemstones and Precious Stones of North America - Leading mining historian Kerby Jackson introduces us to a classic mining publication on the gems and precious stones of the United States, Canada and mexico. First published in 1890, it has been unavailable since those days and sheds important light on the gems and precious stones that may be found in North America. Included are chapters on diamonds, corundum, sapphire, ruby, topaz, emerald, disapore, spinel, turquoise, tourmaline, garnets, beyrl, peridot, zircon, quartz crystals, feldspars, pearls and many others. Included are details on where these gems and precious stones may be found throughout North America, as well as their characteristics. 360 ppgs, 24.99

Mining Camps and Mining Districts - First released in 1885 by Charles Howard Shinn under the title "Mining Camps: A Study in American Frontier Government", this publication offers a unique look at how early gold miners established their own forms of representative government during the California Gold Rush. Drawing on the the early mining codes of mideviel German miners in the Harz Mountains, on the mining customs of the Cornish tin miners and early Spanish mining laws introduced into California, the miners established the first governments in the American West. 340 ppgs, 24.99

BLM Field Handbook for Mineral Examiners - Leading mining historian Kerby Jackson introduces us to a classic mining publication on mine evaluation. First published in 1962, this work sheds important light on the techniques of BLM Mineral Examiners to perform validity on mining claims. 132 ppgs, 10.99

Six Months In The Gold Mines During The California Gold Rush - Unavailable since 1850, this important work is a first hand account of one "49'ers" personal experience during the great California Gold Rush, shedding important light on one of the most exciting periods in the history of not only California, but also the world. Compiled from journals written between 1847 and 1849 by E. Gould Buffum, a native of New York, "Six Months In The Gold Mines During The California Gold Rush" offers a rare look into the day to day lives of the people who came to California to work in her gold mines when the state was still a great frontier. **8.5" X 11", 290 ppgs. Retail Price: $19.99**

The Discovery of Gold in Australia - First published in 1852, it has been unavailable since those days and sheds important light on Australia's gold mining history. Included are rare communications between British agents and the British Crown when gold was first discovered in Australia in 1851. This rare text contains hard to find details on Australia's first mining camps and Britain's early attempts to provide for the orderly regulation of gold mines in that part of the world. Also of interest are hard to find extracts of articles that appeared in the early colonial newspapers that did their best to report on Australia's gold rush as it took place.
102 ppgs, 10.99